Structural Elements Design Manual

TREVOR DRAYCOTT,
CEng, FIStructE

BH NEWNES

In memory of my father Harry Draycott

Newnes
An imprint of Butterworth-Heinemann Ltd
Linacre House, Jordan Hill, Oxford OX2 8DP

℞ A member of the Reed Elsevier Group

OXFORD   LONDON   BOSTON
MUNICH   NEW DELHI   SINGAPORE   SYDNEY
TOKYO   TORONTO   WELLINGTON

First published 1990
Reprinted 1991, 1992, 1994

British Library Cataloguing in Publication Data
Draycott, Trevor
    Structural elements design manual
    1. Structural components. Design
    I. Title
    624.1771

ISBN 0 7506 0313 5

Printed and bound in Great Britain by
Thomson Litho Ltd, East Kilbride, Scotland

Contents

Preface

During twenty years of lecturing on the design of structural elements, I have often been asked to recommend a suitable textbook on the subject. There are several excellent books available dealing individually with the design of structural members in either timber, concrete, masonry or steel. However, books that deal with the design of structural elements in all four materials are scarce. This is particularly so at present, probably because of the changes that have taken place in recent years to British Standards related to structural engineering. It is for this reason that I decided to write this manual.

My primary aim has been to provide a single source of information for students who are new to the topic. In doing so, I have concentrated on the behaviour and practical design of the main elements that comprise a building structure, and have included plenty of worked examples. Therefore the book should prove useful not only to students of structural and civil engineering, but also to those studying for qualifications in architecture, building and surveying, who need to understand the design of structural elements.

The manual is divided into five chapters: general matters, timber, concrete, masonry and steel. Each chapter provides practical guidance on the design of structural elements in accordance with the appropriate British Standard or Code of Practice. However, this manual is not intended to be an exhaustive explanation of the various design codes – although some readers may find it a useful introduction to those codes.

I have been fortunate during thirty years in structural engineering to have received help and guidance, both directly and indirectly, from people too numerous to mention here. This manual is therefore intended, in some small way, to redress the balance by offering assistance to anyone wishing to learn the principles of structural element design. I hope that it will be a source to turn to for help as a student and a friend for reference when one has gained experience and confidence.

In conclusion, I must acknowledge the invaluable assistance received from certain people in particular. I am indebted to Francis Myerscough for the thorough way he read through the draft and for his pertinent comments. Special thanks are due to Sue Dean who somehow managed to decipher the scribble and turn it into a typed manuscript. Last, but certainly not least, I am grateful to my wife and family for their patience and support during the months of writing.

Trevor Draycott

1   General matters

1.1   Introduction

Structural engineering can broadly be described as the study of how the various component elements of a building act together to form a supportive structure and transmit forces down to the foundations. Determining the actual size of the members or elements is only one of the interrelated matters with which the structural engineer is concerned in the design of a building or similar structure. For the purpose of description these matters may be divided into stages and defined as follows:

Structural planning stage   When a structural scheme is devised to suit both the purpose of the building and the site conditions which exist.

Structural analysis stage   When the loads are determined and their dispersal through the structure is analysed by applying the principles of structural mechanics.

Structural elements design stage   When the size needed for each member is calculated in relation to the material and its particular structural capacity.

Structural detailing stage   When detail drawings are produced to illustrate how the structure is to be constructed on site so as to comply with the engineer's design concept

Structural specification stage   When the specification clauses are compiled to ensure that the standard of materials and workmanship to be employed in the works comply with the assumptions embodied in the structural engineer's design.

Building and civil engineering is a team effort, requiring each discipline to have some understanding of the work in others. In this context structural element design is probably the best subject to provide architects, quantity surveyors, building control officers, clerks of works and site staff with a fundamental knowledge of the structural behaviour of the different building materials.

Initially students often mistakenly believe that structural element design is just a form of applied mathematics. Some regrettably are even daunted by this belief. It cannot be denied that in order to determine the size of individual elements it is necessary to carry out calculations, but these, once understood, follow a logical sequence.

To assist us in arriving at a logical design sequence we first need a set of guidelines. These may be found in the relevant British Standards or Codes of Practice which advise on how the materials we use, that is timber, concrete, masonry and steel, behave in the form of building elements such as beams, columns, slabs and walls.

1.2   British Standards   Guidance on the design of building and civil engineering structures is given in various British Standards and Codes of Practice. These play an important role in the provision of structural designs which are both safe and economic and which comply with the Building Regulations and other statutory requirements.

To the inexperienced the standards can be seen as sets of rules restricting freedom and choice, but in the author's opinion they should be accepted as guidelines. Just as our buildings need firm foundations, so too does our knowledge of how structures behave. Engineering judgement and flair come not from taking risks but from a sound understanding of the limits to which we can take the various materials. British Standards contribute to that understanding.

In relation to their application in structural design the various standards and codes may be broadly classified into three groups:

(a) Those relating to the specification of materials and components
(b) Those relating to structural loading
(c) Those relating to the actual design of structural elements in a specific material.

Listed in Tables 1.1, 1.2 and 1.3 respectively are a selection of British Standards within each group.

Table 1.1   Standards relating to materials and components

BSI reference	Title
BS 4 Part 1 1980	Structural steel sections – specification for hot rolled sections
BS 12 1989	Specification for Portland cements
BS 882 1983	Specification for aggregates from natural sources for concrete
BS 890 1972	Specification for building limes
BS 1243 1978	Specification for metal ties for cavity wall construction
BS 3921 1985	Specification for clay bricks
BS 4360 1990	Specification for weldable structural steels
BS 4449 1988	Specification for carbon steel bars for the reinforcement of concrete
BS 4483 1985	Specification for steel fabric for the reinforcement of concrete
BS 4721 1981 (1986)	Specification for ready-mixed building mortars
BS 4978 1988	Specification for softwood grades for structural use
BS 5606 1988	Code of practice for accuracy in building
BS 5977 Part 2 1983	Lintels – specification for prefabricated lintels
BS 6073 Part 1 1981	Specification for precast concrete masonry units
BS 6073 Part 2 1981	Method for specifying precast concrete masonry units
BS 6398 1983	Specification for bitumen damp-proof courses for masonry

Table 1.2   Standards and codes relating to structural loading

BSI reference	Title
BS 648 1964	Schedule of weights of building materials
BS 5977 Part 1 1981 (1986)	Lintels – method for assessment of load
BS 6399 Part 1 1984	Loading for buildings – code of practice for dead and imposed loads
BS 6399 Part 3 1988	Loading for buildings – code of practice for imposed roof loads
CP 3 Chapter V Part 2 1972	Loading – wind loads

Table 1.3   Standards relating to the design of structural elements

BSI reference	Title
BS 5268 Part 2 1988	Structural use of timber – code of practice for permissible stress design, materials and workmanship
BS 5628 Part 1 1978 (1985)	Use of masonry – structural use of unreinforced masonry
BS 5950 Part 1 1990	Structural use of steelwork in building – code of practice for design in simple and continuous construction: hot rolled sections
BS 8110 Part 1 1985	Structural use of concrete – code of practice for design and construction

Extracts from British Standards contained in this manual are reproduced by permission of the British Standards Institution. Complete copies can be obtained from the BSI at Linford Wood, Milton Keynes, MK14 6LE. The BSI also publishes a book entitled *Extracts from British Standards for Students of Structural Design*, which is a compilation of various codes. Since these are in abbreviated form, the publication is intended only as an economic means of assisting students in their studies and not as a substitute for the complete codes.

In conclusion, it should be realized that whilst the theory used in the analysis of structural members (calculation of forces, bending moments and so on) may not change, British Standards and Codes of Practice do. It is therefore essential to ensure that before attempting to design any element of a structure you have the current edition and latest amendments of the relevant standard or code.

1.3 Loading

The actual calculation of the loads supported by individual structural elements is seldom given prominence in textbooks. Therefore, in this section the types of load encountered in structural design are defined and examples illustrating the calculation of such loads are given.

It is not always appreciated that perhaps the most important factor to be considered in the design of a structural member is the assessment of the loads that the member must support or resist. For this to be considered in perspective it must be realized that no matter how accurately the design procedure for a particular member is followed, the member will be either inadequate or uneconomic if the design loads assumed are incorrect.

There are three conditions of loading for which a structural member may have to be designed: dead loading, imposed loading and, when so exposed, wind loading. It is also necessary to consider the effect of combined loads.

Dead loading

This may be defined as the weight of all permanent construction. It will comprise the forces due to the static weights of all walls, partitions, floors, roofs and finishes, together with any other permanent construction.

Dead loads can be calculated from the unit weights given in BS 648 'Schedule of weights of building materials', or from the actual known weight of the materials used if they are of a proprietary type.

The dead load should also include an additional allowance for the weight of the member being designed. Since this cannot be known accurately until the size has been determined, it is necessary initially to estimate the self-weight. This may be checked after the member has been designed and if necessary the design should then be modified accordingly.

Some typical building material weights for use in assessing dead loads, based upon BS 648, are given in Table 1.4.

The unit of force, the newton (N), is derived from the unit of mass, the kilogram (kg), by the relationship that force is equal to mass times the gravitational constant of $9.81 \, \text{m/s}^2$. That is,

$$1000 \, \text{kg} = 1000 \times 9.81 \, \text{kg/s}^2 = 9810 \, \text{N}$$

For structural calculation purposes the load in newtons imposed by the dead weight of the materials may be obtained by multiplying by 10 (strictly 9.81) the kilogram values given in BS 648. For example, if the weight of concrete is $2400 \, \text{kg/m}^3$, then

$$\text{Load imposed} = 2400 \times 10 = 24\,000 \, \text{N/m}^3$$

Alternatively, since the structural engineer usually calculates the load imposed on a structural element in kilonewtons (kN), the tabulated values may be divided by approximately 100. For example, again if the weight of concrete is $2400 \, \text{kg/m}^3$, then

$$\text{Load imposed} = 2400/100 = 24 \, \text{kN/m}^3$$

Table 1.4   Weights of building materials (based on BS 648 1964)

Asphalt		*Lead*	
Roofing 2 layers, 19 mm thick	42 kg/m^2	Sheet, 2.5 mm thick	30 kg/m^2
Damp-proofing, 19 mm thick	41 kg/m^2	*Linoleum* 3 mm thick	6 kg/m^2
Road and footpaths, 19 mm thick	44 kg/m^2	*Plaster*	
Bitumen roofing felts		Two coats gypsum, 13 mm thick	22 kg/m^2
Mineral surfaced bitumen per layer	3.5 kg/m^2	*Plastics sheeting* Corrugated	4.5 kg/m^2
Blockwork		*Plywood*	
Solid per 25 mm thick, stone aggregate	55 kg/m^2	per mm thick	0.7 kg/m^2
Aerated per 25 mm thick	15 kg/m^2	*Reinforced concrete*	2400 kg/m^3
Board		*Rendering*	
Blockboard per 25 mm thick	12.5 kg/m^2	Cement:sand (1:3) 13 mm thick	30 kg/m^2
Brickwork		*Screeding*	
Clay, solid per 25 mm thick medium density	55 kg/m^2	Cement:sand (1:3) 13 mm thick	30 kg/m^2
Concrete, solid per 25 mm thick	59 kg/m^2	*Slate tiles* (depending upon thickness and source)	24–78 kg/m^2
Cast stone	2250 kg/m^3	*Steel*	
Concrete		Solid (mild)	7850 kg/m^3
Natural aggregates	2400 kg/m^3	Corrugated roofing sheets per mm thick	10 kg/m^2
Lightweight aggregates (structural)	1760 kg/m^3 +240 or −160	*Tarmacadam* 25 mm thick	60 kg/m^2
Flagstones		*Terrazzo*	
Concrete, 50 mm thick	120 kg/m^2	25 mm thick	54 kg/m^2
Glass fibre		*Tiling, roof*	
Slab, per 25 mm thick	2.0–5.0 kg/m^2	Clay	70 kg/m^2
Gypsum panels and partitions		*Timber* Softwood	590 kg/m^3
Building panels 75 mm thick	44 kg/m^2	Hardwood	1250 kg/m^3
		Water	1000 kg/m^3
		Woodwool Slabs, 25 mm thick	15 kg/m^2

Imposed loading

This is sometimes termed superimposed loading, live loading or super loading, and may be defined as the loading assumed to be produced by the intended occupancy or use of the structure. It can take the form of distributed, concentrated or impact loads.

BS 6399 Part 1 'Loading for buildings' gives values of imposed load for floors and ceilings of various types of building. Those for residential buildings given in BS 6399 Part 1 Table 5 are reproduced here in Table 1.5.

Part 3 of BS 6399 gives the imposed loads to be adopted for the design of roofs. These consist of snow loading and, where applicable, the loading produced by access on to the roof.

In general for small pitched roof buildings where no access is provided to the roof, other than for routine cleaning and maintenance, a minimum uniformly distributed imposed load of $0.75 \, \text{kN/m}^2$ may be adopted or a concentrated load of $0.9 \, \text{kN}$, whichever produces the worst load effect. A small building in this context must have a width not greater than $10 \, \text{m}$ and a plan area not larger than $200 \, \text{m}^2$, and must have no parapets or other abrupt changes in roof height likely to cause drifting of snow and hence a build-up of load. For situations outside these parameters, reference should be made to BS 6399 Part 3 for the imposed roof load to be adopted.

Wind loading

This may be defined as all the loads acting on a building that are induced by the effect of either wind pressure or wind suction. The pressure exerted by the wind is often one of the most important loads which exposed structures have to resist with regard to overall stability.

CP 3 Chapter V Part 2 'Wind loads' gives the wind speeds to be adopted for the design of buildings relative to their geographical location within the United Kingdom. It also gives pressure coefficients for the various parts of a building, such as roofs and walls, in relation to its size and shape. This code will eventually become Part 2 of BS 6399.

Combined loads

Having obtained individual loading cases, that is dead, imposed and wind, the most onerous combination should be determined and the structure designed accordingly. For a member not exposed to wind, such as a floor beam, this would normally be the combination of dead and imposed loading. For a member exposed to wind, such as the rafter of a truss or portal frame, the combination of dead and imposed load would normally be used to design the member initially. It would then be checked for reversal of stress due to a combination of dead load and wind suction.

Wind loading generally influences the overall stability of a building. Therefore, since the emphasis of this manual is on the design of individual structural elements, only the effects of dead and imposed loads will be examined.

Table 1.5 Imposed loads for residential buildings (BS 6399 Part 1 Table 5)

Floor area usage	Intensity of distributed load (kN/m^2)	Concentrated load (kN)
Type 1: self-contained dwelling units		
All	1.5	1.4
Type 2: apartment houses, boarding houses, lodging houses, guest houses, hostels, residential clubs and communal areas in blocks of flats		
Boiler rooms, motor rooms, fan rooms and the like including the weight of machinery	7.5	4.5
Communal kitchens, laundries	3.0	4.5
Dining rooms, lounges, billiard rooms	2.0	2.7
Toilet rooms	2.0	—
Bedrooms, dormitories	1.5	1.8
Corridors, hallways, stairs, landings, footbridges, etc.	3.0	4.5
Balconies	Same as rooms to which they give access but with a minimum of 3.0	1.5 per metre run concentrated at the outer edge
Cat walks	—	1.0 at 1 m centres
Type 3: hotels and motels		
Boiler rooms, motor rooms, fan rooms and the like, including the weight of machinery	7.5	4.5
Assembly areas without fixed seating,* dance halls	5.0	3.6
Bars	5.0	—
Assembly areas with fixed seating*	4.0	—
Corridors, hallways, stairs, landings, footbridges, etc.	4.0	4.5
Kitchens, laundries	3.0	4.5
Dining rooms, lounges, billiard rooms	2.0	2.7
Bedrooms	2.0	1.8
Toilet rooms	2.0	—
Balconies	Same as rooms to which they give access but with a minimum of 4.0	1.5 per metre run concentrated at the outer edge
Cat walks	—	1.0 at 1 m centres

*Fixed seating is seating where its removal and the use of the space for other purposes is improbable.

Having discussed the types of loading encountered, let us look at some examples. These illustrate how the designer has to convert information about the construction into applied loads on individual structural elements such as beams and columns.

Example 1.1

Timber beams spanning 4 m and spaced at 3 m centres as shown in Figure 1.1 support a timber floor comprising joists and boards together with a plaster ceiling. The load imposed by the dead weight of the floor joists and boards is $0.23 \, kN/m^2$ and by the ceiling $0.22 \, kN/m^2$. If the floor has to support a residential imposed load of $1.5 \, kN/m^2$, calculate the total uniformly distributed load that a single timber floor beam supports.

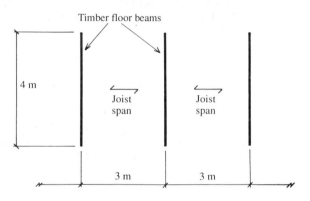

Figure 1.1 *Floor plan*

In this example the structural element under consideration is a timber floor beam. Before a suitable size for this member can be determined the designer must first ascertain the total load it supports. To do so the beam together with the load it carries, which in this instance is a uniformly distributed load (UDL), must be visualized removed from the building (see Figure 1.2).

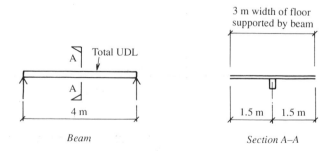

Figure 1.2 *Isolated timber floor beam*

Dead load: joists and boards 0.23
               ceiling <u>0.22</u>
                       $0.45 \, kN/m^2$

Imposed load: 1.5 kN/m²

Combined load: dead   0.45
          imposed   1.5
                   1.95 kN/m²

This combined load is a load per unit area. In order to convert it into a UDL, it must be multiplied by the area supported of 4 m span by 3 m centres. An allowance must also be included for the load due to the self-weight (SW) of the timber beam.

$$\text{Total UDL} = (1.95 \times 4 \times 3) + \text{SW}$$
$$= 23.4 + \text{say } 0.6 = 24.0 \text{ kN}$$

The allowance assumed for the self-weight can be checked after a size for the member has been determined. To illustrate such a check, consider a beam size of 250 mm deep by 100 mm wide and take the average weight of softwood timber as 540 kg/m³.

$$\text{SW} = (540/100) \times 4 \times 0.25 \times 0.1 = 0.54 \text{ kN}$$

Thus the SW of 0.6 assumed was satisfactory.

Example 1.2

Steel floor beams arranged as shown in Figure 1.3 support a 150 mm thick reinforced concrete slab. If the floor has to carry an imposed load of 5 kN/m² and reinforced concrete weighs 2400 kg/m³, calculate the total UDL that each floor beam supports.

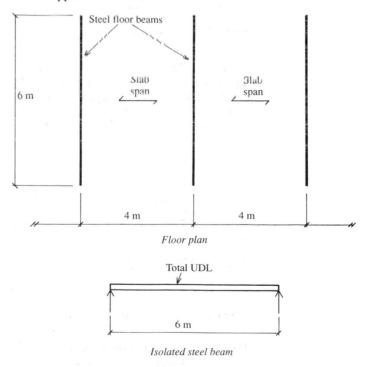

Floor plan

Isolated steel beam

Figure 1.3 *Floor beam arrangement*

The procedure to calculate the UDL is similar to Example 1.1 except that in this case the weight of concrete relates to volume and so needs resolving into a load per unit area to arrive at the dead load of the 150 mm thick slab.

Dead load 150 mm slab: $0.15 \times 2400/100 = 3.6 \, \text{kN/m}^2$

Imposed load: $5 \, \text{kN/m}^2$

Combined load: dead   3.6
            imposed   5.0
                    $8.6 \, \text{kN/m}^2$

$$\text{Total UDL} = (8.6 \times 6 \times 4) + \text{SW}$$
$$= 206.4 + \text{say } 3.6 = 210 \, \text{kN}$$

To check the assumed self-weight, consider that the eventual weight of steel beam will be 60 kg/m run. Then

$$\text{SW} = (60/100) \times 6 = 3.6 \, \text{kN}$$

which is satisfactory.

Example 1.3

Calculate the beam loads and the reactions transmitted to the walls for the steel-work arrangement shown in Figure 1.4. Beam A supports a 100 mm thick reinforced concrete slab, spanning in the direction shown, which carries an imposed load of $3 \, \text{kN/m}^2$. The weight of concrete may be taken as $2400 \, \text{kg/m}^3$ and the weight of the beams as 80 kg/m run.

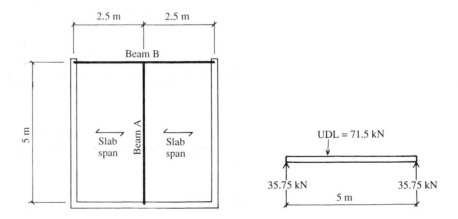

Figure 1.4 *Floor plan*   **Figure 1.5** *Beam A isolated*

Beam A (Figure 1.5) supports a UDL from a 2.5 m width of slab:

Dead load 100 mm slab: $0.1 \times 2400/100 = 2.4 \, \text{kN/m}^2$

Imposed load: $3 \, \text{kN/m}^2$

Combined load: dead   2.4
            imposed   3.0
                    $5.4 \, \text{kN/m}^2$

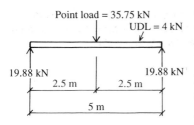

Figure 1.6 *Beam B isolated*

Total UDL = slab UDL + SW

$$= (5.4 \times 5 \times 2.5) + (80/100) \times 5 = 71.5\,\text{kN}$$

Reactions transmitted to wall and beam B $= 71.5/2 = 35.75\,\text{kN}$

Beam B (Figure 1.6) supports the reaction from beam A as a central point load and a UDL due to its self-weight:

Point load $= 35.75\,\text{kN}$

SW UDL $= (80/100) \times 5 = 4\,\text{kN}$

Reactions transmitted to walls $= (35.75 + 4)/2 = 19.88\,\text{kN}$

Example 1.4

A 200 mm thick reinforced concrete mezzanine floor slab is simply supported on beams and columns as shown in Figure 1.7. Calculate the beam and column loads if the floor has to carry an imposed load of $5\,\text{kN/m}^2$. The weight of concrete may be taken as $2400\,\text{kg/m}^3$. The beams may be considered to weigh 120 kg per metre run and the columns 100 kg per metre run.

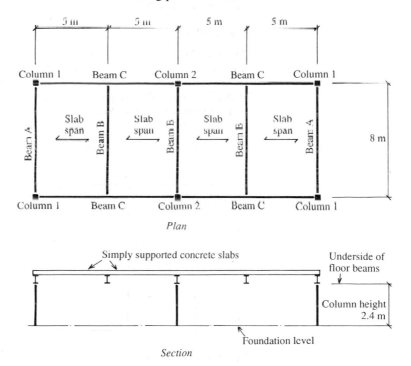

Figure 1.7 *Arrangement of mezzanine floor*

Slab dead load, 200 mm thick: $0.2 \times 2400/100 = 4.8\,\text{kN/m}^2$

Slab imposed load: $5\,\text{kN/m}^2$

Slab combined load: dead   4.8
                   imposed   <u>5.0</u>
                               $9.8\,\text{kN/m}^2$

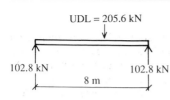

Figure 1.8 *Beam A isolated*

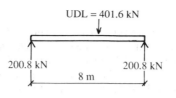

Figure 1.9 *Beam B isolated*

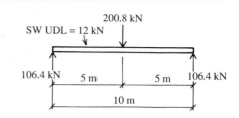

Figure 1.10 *Beam C isolated*

Beam A (Figure 1.8) supports a UDL from a 2.5 m width of simply supported slab together with its SW:

Total UDL = slab UDL + SW UDL
$$= (9.8 \times 8 \times 2.5) + (120/100) \times 8 = 196 + 9.6 = 205.6 \, \text{kN}$$
Reactions = 205.6/2 = 102.8 kN

Beam B (Figure 1.9) supports a UDL from a 5 m width of simply supported slab together with its SW:

Total UDL = slab UDL + SW UDL
$$= (9.8 \times 8 \times 5) + (120/100) \times 8 = 392 + 9.6 = 401.6 \, \text{kN}$$
Reactions = 401.6/2 = 200.8 kN

Beam C (Figure 1.10) supports a central point load from the beam B reaction together with a UDL due to its SW:

SW UDL = (120/100) × 10 = 12 kN
Reactions = (12 + 200.8)/2 = 212.8/2 = 106.4 kN

Figure 1.11 *Column 1 loads*

Column 1 (Figure 1.11) supports the reactions from beam A and C together with its SW:

Beam A reaction	102.8
Beam C reaction	106.4
SW = (100/100) × 2.4	2.4
Total column load	211.6 kN

Column 2 (Figure 1.12) supports the reactions from one beam B and two beams C together with its SW:

Beam B reaction	200.8
Beam C reaction	106.4
Beam C reaction	106.4
SW = (100/100) × 2.4	2.4
	416.0 kN

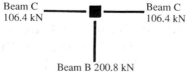

Figure 1.12 *Column 2 loads*

Example 1.5

A series of reinforced concrete beams at 5 m centres span 7.5 m on to reinforced concrete columns 3.5 m high as shown in Figure 1.13. The beams, which are 400 mm deep by 250 mm wide, carry a 175 mm thick reinforced concrete simply supported slab, and the columns are 250 mm by 250 mm in cross-section. If the floor imposed loading is 3 kN/m² and the weight of reinforced concrete is 2400 kg/m³, calculate: the total UDL carried by a beam; the reactions transmitted to the columns; and the load transmitted to the column foundations.

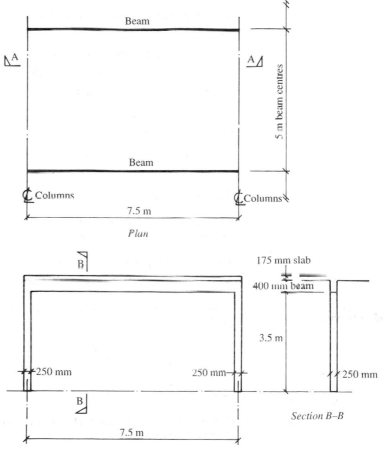

Figure 1.13 *Arrangement of beams and columns*

Combined floor loading: slab SW dead = (2400/100) × 0.175 = 4.2
imposed   = 3.0
                                    7.2 kN/m²

Beam total UDL = UDL from slab + beam SW UDL
        = (7.2 × 7.5 × 5) + (2400/100) × 7.5 × 0.4 × 0.25
        = 270 + 18 = 288 kN

Beam reactions transmitted to columns = (total UDL)/2 = 288/2 = 144 kN

Column foundation load = beam reaction + column SW

$$= 144 + (2400/100) \times 3.5 \times 0.25 \times 0.25$$
$$= 144 + 5.25 = 149.25 \, kN$$

1.4 Structural mechanics

Before the size of a structural element can be determined it is first necessary to know the forces, shears, bending moments and so on that act on that element. It is also necessary to know what influence these have on the stability of the element and how they can be resisted. Such information is obtained by reference to the principles of structural mechanics.

The reader should already be familiar with the principles of structural mechanics. However, two particular topics play a sufficiently important part in design to be repeated here. They are the theory of bending and the behaviour of compression members.

1.5 Theory of bending

The basic design procedure for beams conforms to a similar sequence irrespective of the beam material, and may be itemized as follows:

(a) Calculate the applied loads including the reactions and shear forces.
(b) Calculate the externally applied bending moments induced by the applied loads.
(c) Design the beam to resist the loads, shears, bending moments and resulting deflection in accordance with the guidelines appertaining to the particular beam material.

Item (a) allows a load diagram to be produced and also enables a shear force (SF) diagram to be drawn from which the maximum shear force can be determined.

The induced bending moments, item (b), can be derived from the SF diagram together with the location and magnitude of the maximum bending moment. This coincides with the point of zero shear, which is also known as the point of contraflexure. A bending moment (BM) diagram can then be drawn.

Formulae are given in various design manuals for calculating the maximum bending moments and deflections of simply supported beams carrying standard loading patterns such as a central point load, or equally spaced point loads, or a uniformly distributed load. The loading, shear force, bending moment and deflection diagrams for the two most common load conditions are illustrated together with the relevant formulae in Figure 1.14. That for a constant uniformly distributed load (UDL) is shown in Figure 1.14a and that for a central point load in Figure 1.14b. For unsymmetrical loading patterns the reactions, shear force, bending moment and deflection values have to be calculated from first principles using the laws of basic statics.

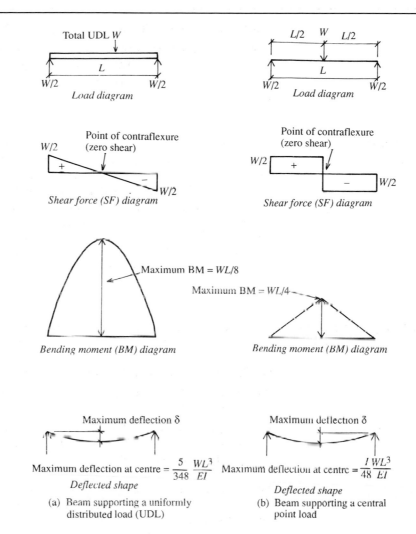

Figure 1.14 *Load, shear force and bending moment diagrams for standard loading conditions*

The resistance of a beam to bending, referred to in item (c), is derived from the theory of bending. The general expression for the theory of bending is

$$\frac{M}{I} = \frac{f}{y} = \frac{E}{R}$$

where

M   either the internal moment of resistance (MR) of the beam or the external bending moment (BM) applied to the beam

I   second moment of area of the beam which is a geometrical property of the beam

f stress value for the beam (dependent on the beam material, such as timber or steel)

y distance from the neutral axis (NA) of the beam to its extreme fibres

E Young's modulus of elasticity for the beam (again dependent on the beam material)

R radius of curvature after bending

The term E/R relates to the deformation of a beam and is used in the derivation of deflection formulae. It is not used in bending calculations, and the expression therefore reduces to

$$\frac{M}{I} = \frac{f}{y}$$

This expression may be rearranged so that

$$M = f\frac{I}{y} \quad \text{or} \quad f = M\frac{y}{I}$$

Now I/y is a geometric property of a beam section called the elastic modulus or section modulus, and is denoted by the symbol Z. Thus

$$M = fZ \tag{1.1}$$

or

$$f = \frac{M}{Z} \tag{1.2}$$

or

$$Z = \frac{M}{f} \tag{1.3}$$

The equations can be used in design as follows:

(a) Equation 1.1 may be used to calculate the internal moment of resistance (MR) for a beam of known size (Z known) and material (f known).

(b) Equation 1.2 may be used to calculate the bending stress f occurring within a beam of known size (Z known) when it is subjected to an externally applied bending moment (BM known).

(c) Equation 1.3 may be used for a beam of known material (f known) to calculate the beam property Z needed for the beam to resist an externally applied bending moment (BM known).

The key to their use is the relationship between a beam's moment of resistance (MR) and the applied bending moment (BM). If a beam section is not to fail under load, an internal moment of resistance (MR) must be developed within the beam at least equal to the maximum external bending moment (BM) produced by the loads. That is,

Internal MR = external BM

Consider the simply supported rectangular beam shown in Figure 1.15a. If a load were applied to the beam it would bend as shown exaggerated in Figure 1.15b. The deformation that takes place in bending causes the fibres above the neutral axis (NA) of the beam to shorten or compress and those below to stretch. In resisting this shortening and stretching the fibres of the beam are placed in compression and tension respectively. This induces compressive stresses above the NA and tensile stresses below. These are a maximum at the extreme fibres and zero at the NA, as indicated on the Figure 1.15b stress diagram.

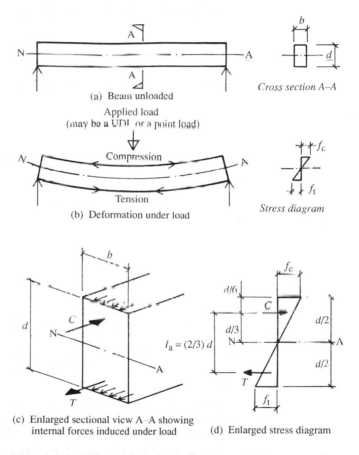

(a) Beam unloaded

Cross section A–A

(b) Deformation under load

Stress diagram

(c) Enlarged sectional view A–A showing internal forces induced under load

(d) Enlarged stress diagram

Figure 1.15 *Theory of bending related to a simply supported rectangular beam*

By reference to the enlarged beam cross-section Figure 1.15c and stress diagram Figure 1.15d, it can be seen that a couple is set up within the beam comprising a compressive force C and a tensile force T acting at the centres of gravity of the stress blocks with a lever arm of $(2/3)d$. The moment of resistance of the beam is the product of this couple:

$$MR = \text{force } C \text{ (or } T) \times \text{lever arm}$$
$$= \text{force } C \text{ (or } T) \times \tfrac{2}{3}d$$

The forces C and T are equal to the average stress multiplied by the surface area upon which it acts. The average stress is either $f_c/2$ or $f_t/2$; the surface area is half the beam cross-section, that is $bd/2$. Therefore

$$\text{Force } C = \frac{f_c}{2} \times \frac{bd}{2} = f_c \frac{bd}{4}$$

$$\text{Force } T = \frac{f_t}{2} \times \frac{bd}{2} = f_t \frac{bd}{4}$$

hence

$$\text{MR} = f_c \frac{bd}{2} \times \frac{2}{3}d \quad \text{or} \quad f_t \frac{bd}{4} \times \frac{2}{3}d$$

$$= f_c \frac{bd^2}{6} \qquad \text{or} \quad f_t \frac{bd^2}{6}$$

Since the values of f_c and f_t are the same, the symbol f may be adopted for the stress. So

$$\text{MR} = f \frac{bd^2}{6}$$

The term $bd^2/6$ is the section modulus Z, referred to earlier, for a rectangular beam of one material such as timber. Rectangular reinforced concrete beams are composite beams consisting of two materials and are therefore not within this category.

For rolled steel beams a value for the section or elastic modulus is obtained directly from steel section property tables. The values for the section modulus Z are given in length units[3] and those for the second moment of area in length units[4].

Let us now examine the use of the theory of bending for simple beam design.

Example 1.6

A timber beam spanning 4.5 m supports a UDL of 4 kN including its self-weight, as shown in Figure 1.16. Assuming the breadth of the beam to be 50 mm and the allowable stress in timber to be 7 N/mm², what depth of beam is required?

We have $b = 50$ mm and $f = 7$ N/mm²; d is to be found. First,

$$\text{Internal MR} = \text{external BM maximum}$$

$$\frac{fbd^2}{6} = \frac{WL}{8}$$

now

$$\frac{fbd^2}{6} = \frac{7 \times 50 \times d^2}{6} \, \text{N mm}$$

$$\frac{WL}{8} = \frac{4 \times 4.5}{8} \, \text{kN m}$$

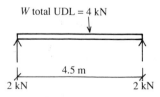

W total UDL = 4 kN

4.5 m

2 kN   2 kN

Figure 1.16 *Load diagram*

It is necessary to make the units compatible on both sides of the relationship. Let us multiply the BM units by 10^3 to convert the kilonewtons to newtons, and by a further 10^3 to convert metres to millimetres. Then

$$\frac{7 \times 50 \times d^2}{6} = \frac{4 \times 10^3 \times 4.5 \times 10^3}{8}$$

$$d^2 = \frac{4 \times 4.5 \times 10^6 \times 6}{8 \times 7 \times 50}$$

$$d = \sqrt{\left(\frac{4 \times 4.5 \times 10^6 \times 6}{8 \times 7 \times 50} \right)} = 196.4 \,\text{mm}$$

Use a 50 mm × 200 mm timber beam.

Example 1.7

Calculate the depth required for the timber beam shown in Figure 1.17a if the breadth is 75 mm and the permissible bending stress is 8.5 N/mm². An allowance for the self-weight of the beam has been included with the point loads.

We have $b = 75$ mm and $f = 8.5$ N/mm²; d is to be found. To complete the load diagram it is first necessary to calculate the reactions. Take moments about end B, clockwise moments being positive and anti-clockwise moments negative:

$$8R_a = (3 \times 6) + (5 \times 2)$$
$$8R_a = 18 + 10$$
$$8R_a = 28$$
$$R_a = 28/8 = 3.5 \,\text{kN}$$

Therefore $R_b = 8 - 3.5 = 4.5$ kN.

Having calculated the reactions to complete the load diagram, the shear force diagram Figure 1.17b can be constructed. This shows that a point of contraflexure occurs under the 5 kN point load, and hence the maximum bending moment will be developed at that position. The bending moment diagram for the beam is shown in Figure 1.17c.

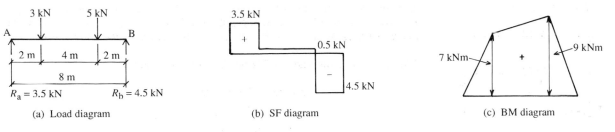

3 kN   5 kN

A   B

2 m   4 m   2 m

8 m

$R_a = 3.5$ kN $R_b = 4.5$ kN

(a) Load diagram

3.5 kN

0.5 kN

4.5 kN

(b) SF diagram

7 kNm   9 kNm

(c) BM diagram

Figure 1.17 *Timber beam diagrams*

BM under 3 kN point load $= 3.5 \times 2 = 7 \,\text{kN m} = 7 \times 10^6 \,\text{N mm}$

Maximum BM under 5 kN point load $= 4.5 \times 2 = 9 \,\text{kN m} = 9 \times 10^6 \,\text{N mm}$

The maximum BM is equated to the internal moment of resistance:

$$\text{Internal MR} = \text{external BM maximum}$$

$$f\frac{bd^2}{6} = 9 \times 10^6$$

$$\frac{8.5 \times 75 \times d^2}{6} = 9 \times 10^6$$

$$d = \sqrt{\left(\frac{9 \times 10^6 \times 6}{8.5 \times 75}\right)} = 291 \text{ mm}$$

Use a 75 mm × 300 mm timber beam.

Example 1.8

A steel beam supports a total UDL including its self-weight of 65 kN over a span of 5 m. If the permissible bending stress for this beam is taken as 165 N/mm², determine the elastic modulus needed for the beam.

We have

$$\text{Internal MR} = \text{external BM maximum}$$

$$fZ = \frac{WL}{8}$$

$$165Z = \frac{65 \times 10^3 \times 5 \times 10^3}{8}$$

$$Z = \frac{65 \times 5 \times 10^6}{8 \times 165}$$

$$Z = 246\,212 \text{ mm}^3 = 246.21 \text{ cm}^3$$

Therefore the elastic modulus Z needed for the beam is 246.21 cm³. Section property tables for steel beams give the elastic modulus values in cm³ units. By reference to such tables we see that a 254 mm × 102 mm × 25 kg/m universal beam section, which has an elastic modulus of 265 cm³, would be suitable in this instance.

Example 1.9

A timber beam spanning 5 m supports a UDL of 4 kN which includes an allowance for its self-weight. If a 100 mm wide by 200 mm deep beam is used, calculate the bending stress induced in the timber. What amount of deflection will be produced by the load if the E value for the timber is 6600 N/mm², and how does this compare with a permissible limit of 0.003 × span?

We know $b = 100$ mm and $d = 200$ mm; f is to be found. We have

$$\text{Internal MR} = \text{external BM maximum}$$

$$f\frac{bd^2}{6} = \frac{WL}{8}$$

$$\frac{f \times 100 \times 200^2}{6} = \frac{4 \times 10^3 \times 5 \times 10^3}{8}$$

$$f = \frac{4 \times 5 \times 10^6 \times 6}{8 \times 100 \times 200^2} = 3.75 \text{ N/mm}^2$$

The actual deflection is given by

$$\delta_a = \frac{5}{384} \frac{WL^3}{EI}$$

The second moment of area I for a rectangular section is given by

$$I = \frac{bd^3}{12} = \frac{100 \times 200^3}{12} = 66.66 \times 10^6 \, \text{mm}^4$$

Therefore

$$\delta_a = \frac{5}{384} \times \frac{4 \times 10^3 \times 5000^3}{6600 \times 66.66 \times 10^6} = 14.8 \, \text{mm}$$

The permissible deflection is given by

$$\delta_p = 0.003 \times \text{span} = 0.003 \times 5000 = 15 \, \text{mm}$$

The actual deflection of 14.8 mm is less than the permissible 15 mm and therefore the beam would be adequate in deflection.

1.6 Compression members

Compression members are those elements within a structure which have to resist compressive stresses induced by the loads they support. The most obvious examples to be found in a building are the main vertical support members to the roof and floors. These are commonly referred to as either columns, posts or stanchions depending on the material from which they are formed. Reinforced concrete compression members are usually called columns, timber compression members posts, and steel compression members stanchions.

The vertical loads they support can be concentric or eccentric. If the load is concentric its line of application coincides with the neutral axis (NA) of the member (see Figure 1.18). Such compression members are said to be axially loaded and the stress induced is a direct compressive stress. In practice the vertical load is often applied eccentrically so that its line of action is eccentric to the NA of the member (see Figure 1.19). This induces compressive bending stresses in the member in addition to direct compressive stresses.

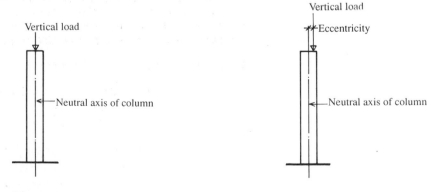

Figure 1.18 *Concentric load*   **Figure 1.19** *Eccentric load*

In relation to their mode of failure, compression members can be described as either 'short' or 'long'. A short compression member would fail due to the material crushing, whereas a long member may fail by buckling laterally before crushing failure of the material is reached (see Figure 1.20).

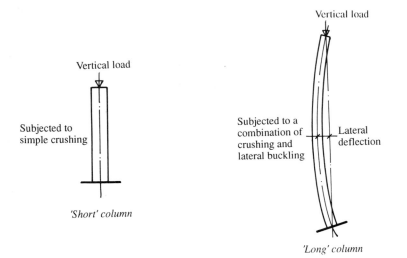

Figure 1.20 *Short and long columns*

A reinforced concrete column is considered to be a short column when its effective height does not exceed fifteen times its least width. Thus a column of 300 mm by 200 mm cross-section is in the short category when its effective height does not exceed 3 m. A large majority of reinforced concrete columns are in this category.

The design of axially loaded short columns is simply based on the expression

$$\text{Stress} = \frac{\text{load}}{\text{area}}$$

where the stress is the permissible compressive stress of the column material, the load is the applied vertical load, and the area is the cross-sectional area of the column.

Steel sections are produced by rolling the steel whilst hot into various standard cross-sectional profiles. Some of the typical shapes available are shown in Figure 1.21. Information on the dimensions and geometric properties of standard steel sections may be obtained from British Standards or publications produced by the Steel Construction Institute.

For a steel column to be considered as a short column, its effective height must generally not exceed six times its least width. Hence, the effective height of a 203 mm × 203 mm universal column (UC) section would not have to exceed 1218 mm for it to be a short column. This serves to illustrate that in practical terms steel columns are usually in the long column category.

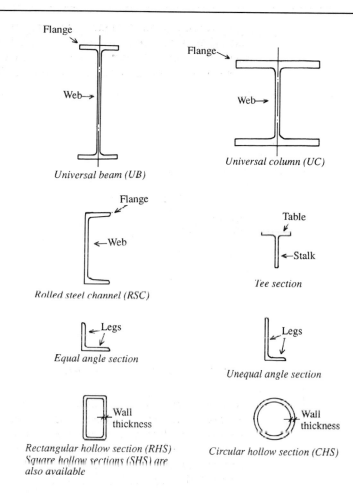

Figure 1.21 *Rolled steel sections, typical shapes available*

Timber posts, because of the cross-sectional dimensions of available sizes, are normally also in the long column category.

Since long columns fail due to a combination of crushing and lateral buckling, the permissible stress is related to their slenderness. This depends upon the column height, its cross-sectional geometry and how it is held at the top and bottom.

The factor which governs the permissible stress of a long column is its slenderness ratio. This is the ratio of the effective length to the least radius of gyration of the member. The permissible compressive stress reduces as the slenderness ratio of the column increases. Thus

$$\text{Slenderness ratio} = \frac{\text{effective length}}{\text{least radius of gyration}}$$

$$SR = \frac{l}{r}$$

The radius of gyration is another geometric property, related to the second moment of area of the column section and its area:

$$\text{Radius of gyration} = \sqrt{\left(\frac{\text{second moment of area}}{\text{area}}\right)}$$

$$r = \sqrt{\left(\frac{I}{A}\right)}$$

The effective length of a column is controlled by the way it is held at each end or, as it is termed, its end fixity. By effective length we mean the height of column which is subject to lateral buckling.

If a column is located in position at each end but not held rigidly, it would buckle over a distance equivalent to its full height (see Figure 1.22). The ends in this instance are said to be held in position only or pinned. If the column were held rigidly at each end, however, the distance over which it would tend to buckle would reduce to something less than its full height (see Figure 1.23). In this instance the ends are said to be both held in position and restrained in direction or fixed.

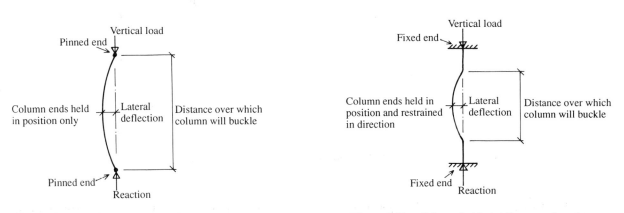

Figure 1.22   *Column located in position only at each end*   **Figure 1.23**   *Column held rigidly at each end*

It can be seen therefore that different conditions of end fixity produce different effective lengths. In general there are four standard effective length conditions; these are illustrated in Figure 1.24. Guidance on the type of end connection needed to produce the different end restraint conditions in relation to the various materials is given in the relevant British Standards. It should be understood that, all other things being equal, the shorter the effective length the stronger the member.

There are subtle differences in the design approach for columns depending on the material. Therefore, to avoid confusion, examples on the design of columns will be dealt with in each of the respective material chapters of this manual.

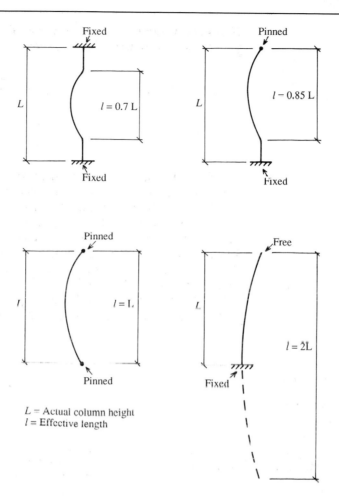

Figure 1.24 *Effective length conditions for columns*

1.7 Summary

There are few aspects of structural design that do not benefit from the adoption of a methodical procedure to minimize the chance of error.

In relation to the general matters dealt with in this chapter, these may be summarized into the following procedural list:

(a) Evaluate the loads acting on the structure.
(b) Determine the loads acting on the individual structural members.
(c) Calculate the forces, shears, bending moments and so on induced in each member by the loads.
(d) Design the respective members.

Step (d) depends on the design guidelines for the particular material from which the members are formed. The reader should therefore refer to the relevant chapter of this manual for the design of structural elements in a specific material.

2   Timber elements

2.1   Stress grading

Of all the materials used for construction, timber is unique by virtue of being entirely natural. Whilst this gives it a deserved aesthetic appeal, it also creates an initial problem for the structural engineer.

In order to design any structural component efficiently, it is necessary to know in advance the strength capability of the material to be used. Timber presents a problem in this respect since we have no apparent control over its quality. All the other materials we use structurally are man made and therefore some form of quality control can be exercised during their production.

To overcome this difficulty and to enable timber to compete equally with other structural materials, the stress grading method of strength classification has been devised. This is based on an assessment of features in timber that are known to influence strength. Guidance for such assessment either by visual inspection or by use of stress grading machines is given for softwoods in BS 4978 'Specification for softwood grades for structural use'. The implications of this code will be discussed in more detail here. For guidance on the stress grading of tropical hardwoods reference should be made to BS 5756 'Specification for tropical hardwoods graded for structural use'.

Visual stress grading is a manual process carried out by approved graders who have been trained and have demonstrated their proficiency in the technique. The grader examines each piece of timber to check the size and frequency of specific physical characteristics: knots, slope of grain, rate of growth, wane, resin pockets and distortion. These are compared with the permitted limits given in BS 4978 to determine whether a piece is accepted into one of the two visual stress grades or rejected. The two visual grades referred to in the standard are general structural (GS) grade and special structural (SS) grade.

The machine stress grading method is based on the principle that strength is related to stiffness. Therefore, since stiffness may be established by measuring deflection under load, the method offers the basis for a non-destructive testing technique. Stress grading machines employ such a technique. Timber is passed through the machine and, by means of a series of rollers, some static and some exerting pressure, bending is induced at increments along its length. The resulting deflection is measured by a computer linked to the machine and compared simultaneously with pre-programmed parameters for accepting or rejecting the timber into one of four machine grades.

The four machine grades specified in BS 4978 are MGS, MSS, M50 and M75. In order that stress graded timber may be identified, every piece is indelibly marked, on at least one face, with its grade and the company or machine which graded it.

No design stresses are actually given in BS 4978; the code simply provides grading rules to enable timber suppliers to categorize timber. Reference should be made to BS 5268 for the relevant grade stresses to be adopted and for guidance on various aspects that should be considered in the structural design of timber elements.

2.2 Structural design of timber

Guidance on the use of timber in building and civil engineering structures is given in BS 5268 'Structural use of timber'. This is divided into the following seven parts:

Part 1   Limit state design, materials and workmanship.

Part 2   Code of practice for permissible stress design, materials and workmanship.

Part 3   Code of practice for trussed rafter roofs.

Part 4   Fire resistance of timber structures.

Part 5   Preservation treatments for constructional timber.

Part 6   Code of practice for timber frame walls.

Part 7   Recommendations for the calculation basis for span tables.

The structural design of timber members in this manual will be related to Part 2 of the standard, which is based on permissible stress philosophy. This follows the principles of elastic behaviour, from which are derived both the theory of bending and the behaviour of compression members that were discussed in Chapter 1.

2.3 Symbols

Those symbols used in BS 5268 that are relevant to this manual are as follows:

Bending

BM, M bending moment

$\sigma_{m,a,par}$ applied bending stress parallel to grain

$\sigma_{m,g,par}$ grade bending stress parallel to grain

$\sigma_{m,adm,par}$ permissible bending stress parallel to grain

Shear

F_v vertical external shear force

r_a applied shear stress parallel to grain

r_g grade shear stress parallel to grain

r_{adm} permissible shear stress parallel to grain

Deflection

δ_p permissible deflection

δ_m bending deflection

δ_v shear deflection

E modulus of elasticity

E_{mean} mean value of E

E_{min} minimum value of E

G shear modulus (modulus of rigidity)

Section properties

A total cross-sectional area

b breadth

h depth of a beam

h_{e} effective depth of a beam

i radius of gyration

I second moment of area

L length, span

L_{e} effective length of a column

$\lambda = L_{\text{e}}/i$ slenderness ratio (expressed in terms of radius of gyration)

$\lambda = L_{\text{e}}/b$ slenderness ratio (expressed in terms of breadth of section)

Z section modulus

Compression

$\sigma_{\text{c,a,par}}$ applied compression stress parallel to grain

$\sigma_{\text{c,g,par}}$ grade compression stress parallel to grain

$\sigma_{\text{c,adm,par}}$ permissible compression stress parallel to grain

2.4 Strength classes

By reference to BS 5268 Part 2, timber that has been categorized by stress grading may be further classified into strength classes in relation to the grade and species of the timber. There are nine strength classes from the weakest, lowest grade, softwood SC1 to the densest, highest grade, hardwood SC9. Softwoods are covered by classes SC1 to SC5, and hardwoods by classes SC6 to SC9.

The various timber species are assigned into strength classes by Tables 3–8 of BS 5268. Table 3, which is that for softwood species and grade combinations graded in accordance with BS 4978, is reproduced here as Table 2.1. Tables 4, 5, 6 and 7 relate to North American timbers and Table 8 to tropical hardwoods.

It is possible for stress grading machines to be set to allot timber directly into strength classes. Timber graded in this way would be marked with the relevant strength class reference SC1, SC2 and so on.

Strength class classification is intended to simplify the design, specification and supply of structural timber.

2.5 Grade stresses

Grade stresses for each of the nine strength classes are given, without reference to timber species, in Table 9 of BS 5268. This is reproduced here as Table 2.2. By choosing one of the strength classes from the table, the designer can determine the size of a timber member without specifying its species. The supplier may then provide any species from within the stipulated strength class.

Table 2.1 Softwood species/grade* combinations which satisfy the requirements for strength classes: graded to BS 4978 (BS 5268 Part 2 1988 Table 3)

Standard name	Strength class†				
	SC1	SC2	SC3	SC4	SC5‡
Imported					
Parana pine			GS	SS	
Pitch pine (Caribbean)			GS		SS
Redwood			GS/M50	SS	M75
Whitewood			GS/M50	SS	M75
Western red cedar	GS	SS			
Douglas fir-larch (Canada)			GS	SS	
Douglas fir-larch (USA)			GS	SS	
Hem-fir (Canada)			GS/M50	SS	M75
Hem-fir (USA)§			GS	SS	
Spruce-pine-fir (Canada)§			GS/M50	SS/M75	
Sitka spruce (Canada)¶		GS	SS		
Western whitewoods (USA)¶	GS		SS		
Southern pine (USA)			GS	SS	
British grown					
Douglas fir		GS	M50/SS		M75
Larch			GS	SS	
Scots pine			GS/M50	SS	M75
Corsican pine		GS	M50	SS	M75
European spruce¶	GS	M50/SS	M75		
Sitka spruce¶	GS	M50/SS	M75		

* Machine grades MGS and MSS are interchangeable with GS and SS grades respectively. The S6, S8, MS6 and MS8 grades of the ECE 'Recommended standard for stress grading of coniferous sawn timber' (1982) may be substituted for GS, SS, MGS and MSS respectively

† A species/grade combination from a higher strength class (see Table 9 of BS 5268, here Table 2.2) may be used where a lower strength class is specified.

‡ All softwoods classified as or machine graded to strength class SC5, except pitch pine and southern pine (USA), should use the fastener loads tabulated for strength classes SC3 and SC4.

§ For grades of hem-fir (USA) and spruce-pine-fir (Canada) classified as or machine graded to strength classes other than SC1 and SC2, the values of lateral load perpendicular to the grain for bolts and timber connectors should be multiplied by the joint/class modification factor K_{42} which has the value 0.9.

¶ All grades of British grown Sitka spruce, Canadian Sitka spruce, British grown European spruce and western whitewoods (USA) should use the fastener loads tabulated for strength classes SC1 and SC2.

It may sometimes be necessary to specify a particular species for reasons other than strength. This may be for appearance, durability or other material quality. In such instances the designer, having designed to a particular strength class, can specify the species from within that class which he is prepared to accept.

Table 2.2   Grade stresses and moduli of elasticity for strength classes: for the dry exposure condition (BS 5268 Part 2 1988 Table 9)

Strength class	Bending parallel to grain	Tension parallel to grain	Compression parallel to grain	Compression perpendicular to grain*		Shear parallel to grain	Modulus of elasticity		Approximate density†
							Mean	Minimum	
	(N/mm^2)	(N/mm^2)	(N/mm^2)	(N/mm^2)	(N/mm^2)	(N/mm^2)	(N/mm^2)	(N/mm^2)	(kg/m^3)
SC1	2.8	2.2‡	3.5	2.1	1.2	0.46	6 800	4 500	540
SC2	4.1	2.5‡	5.3	2.1	1.6	0.66	8 000	5 000	540
SC3	5.3	3.2‡	6.8	2.2	1.7	0.67	8 800	5 800	540
SC4	7.5	4.5‡	7.9	2.4	1.9	0.71	9 900	6 600	590
SC5	10.0	6.0‡	8.7	2.8	2.4	1.00	10 700	7 100	590/760
SC6§	12.5	7.5	12.5	3.8	2.8	1.50	14 100	11 800	840
SC7§	15.0	9.0	14.5	4.4	3.3	1.75	16 200	13 600	960
SC8§	17.5	10.5	16.5	5.2	3.9	2.00	18 700	15 600	1080
SC9§	20.5	12.3	19.5	6.1	4.6	2.25	21 600	18 000	1200

*When the specification specifically prohibits wane at bearing areas, the higher values of compression perpendicular to the grain stress may be used; otherwise the lower values apply.

†Since many species may contribute to any of the strength classes, the values of density given in this table may be considered only crude approximations. When a more accurate value is required it may be necessary to identify individual species and utilize the values given in Appendix A of BS 5268. The higher value for SC5 is more appropriate for hardwoods.

‡Note the light framing, stud, structural light framing no. 3 and joist and plank no. 3 grades should not be used in tension.

§Classes SC6, SC7, SC8 and SC9 will usually comprise the denser hardwoods.

2.6   Design stresses

The grade stresses given in Table 2.2 are basic stresses applicable to timber in a dry exposure condition, within certain dimensional and geometrical parameters and subjected to permanent loading. If any of these conditions change then the basic grade stress is affected. Therefore, to obtain the permissible design stresses, modification factors known as K factors are given in BS 5268 to be used when necessary to adjust the grade stress.

A summary of the K factors which are relevant to the designs contained in this manual are as follows:

K_1 wet exposure geometrical property modification factor

K_2 wet exposure stress modification factor

K_3 load duration modification factor

K_5 notched end shear stress modification factor

K_7 bending stress, depth modification factor

K_8 load-sharing modification factor (= 1.1)

K_{12} slenderness ratio modification factor for compression members

The use of these factors will be discussed further in the relevant sections of this chapter.

2.7 Dry and wet exposure conditions

The terms 'dry' and 'wet' refer to the exposure conditions that will exist when the timber is in service, and relate to the moisture content of such timber. Hence dry exposure timber will have an average moisture content not exceeding 18 per cent. This exposure includes most covered buildings and internal uses. Wet exposure timber will have an average moisture content greater than 18 per cent. Such timber would generally be in an external environment or in contact with water.

The grade stresses given in Table 2.2 are for timber in the dry exposure condition. These are reduced in the wet state by multiplying by the appropriate K_2 factor from BS 5268 Table 16, reproduced here as Table 2.3.

Table 2.3 Modification factor K_2 by which dry stresses and moduli should be multiplied to obtain wet stresses and moduli applicable to wet exposure conditions (BS 5268 Part 2 1988 Table 16)

Property	Value of K_2
Bending parallel to grain	0.8
Tension parallel to grain	0.8
Compression parallel to grain	0.6
Compression perpendicular to grain	0.6
Shear parallel to grain	0.9
Mean and minimum modulus of elasticity	0.8

2.8 Geometrical properties of timber.

The geometrical properties for sawn, planed all round and regularized softwoods are given in Tables 98, 99 and 100 respectively of BS 5268. That for sawn softwoods is reproduced here as Table 2.4.

The tables relate to timber in the dry exposure condition. For timber in the wet exposure condition the values contained in the tables should be multiplied by the relevant K_1 factor, obtained from Table 2 of BS 5268 and reproduced here as Table 2.5. This has the effect of increasing the geometrical properties induced by swelling of the timber when wet.

2.9 Duration of load

The grade stresses given in Table 2.2 are applicable to timber members supporting permanent loads. However, a timber member can support a greater load for a short period than it can permanently. This fact is allowed for in design by adjusting the grade stresses using an appropriate load duration factor K_3 from BS 5268 Table 17, which is reproduced here as Table 2.6.

2.10 Load sharing systems

A load sharing system is said to exist when a minimum of four members, such as rafters, joists, trusses or wall studs, are placed at centres not exceeding 610 mm and adequate provision is made for the lateral distribution of the applied loads via purlins, binders, boards or battens.

In such instances, a 10 per cent increase in the appropriate grade stress

Table 2.4   Geometrical properties of sawn softwoods (BS 5268 Part 2 1988 Table 98)

Basic size* (mm)	Area (10^3 mm^2)	Section modulus About x–x (10^3 mm^3)	About y–y (10^3 mm^3)	Second moment of area About x–x (10^6 mm^4)	About y–y (10^6 mm^4)	Radius of gyration About x–x (mm)	About y–y (mm)
36 × 75	2.70	33.8	16.2	1.27	0.292	21.7	10.4
36 × 100	3.60	60.0	21.6	3.00	0.389	28.9	10.4
36 × 125	4.50	93.8	27.0	5.86	0.486	36.1	10.4
36 × 150	5.40	135	32.4	10.1	0.583	43.3	10.4
38 × 75	2.85	35.6	18.1	1.34	0.343	21.7	11.0
38 × 100	3.80	63.3	24.1	3.17	0.457	28.9	11.0
38 × 125	4.75	99.0	30.1	6.18	0.572	36.1	11.0
38 × 150	5.70	143	36.1	10.7	0.686	43.3	11.0
38 × 175	6.65	194	42.1	17.0	0.800	50.5	11.0
38 × 200	7.60	253	48.1	25.3	0.915	57.7	11.0
38 × 225	8.55	321	54.2	36.1	1.03	65.0	11.0
44 × 75	3.30	41.3	24.2	1.55	0.532	21.7	12.7
44 × 100	4.40	73.3	32.3	3.67	0.710	28.9	12.7
44 × 125	5.50	115	40.3	7.16	0.887	36.1	12.7
44 × 150	6.60	165	48.4	12.4	1.06	43.3	12.7
44 × 175	7.70	225	56.5	19.7	1.24	50.5	12.7
44 × 200	8.80	293	64.5	29.3	1.42	57.7	12.7
44 × 225	9.90	371	72.6	41.8	1.60	65.0	12.7
44 × 250	11.0	458	80.7	57.3	1.77	72.2	12.7
44 × 300	13.2	660	96.8	99.0	2.13	86.6	12.7
47 × 75	3.53	44.1	27.6	1.65	0.649	21.7	13.6
47 × 100	4.70	78.3	36.8	3.92	0.865	28.9	13.6
47 × 125	5.88	122	46.0	7.65	1.08	36.1	13.6
47 × 150	7.05	176	55.2	13.2	1.30	43.3	13.6
47 × 175	8.23	240	64.4	21.0	1.51	50.5	13.6
47 × 200	9.40	313	73.6	31.3	1.73	57.7	13.6
47 × 225	10.6	397	82.8	44.6	1.95	65.0	13.6
47 × 250	11.8	490	92.0	61.2	2.16	72.2	13.6
47 × 300	14.1	705	110	106	2.60	86.6	13.6
50 × 75	3.75	46.9	31.3	1.76	0.781	21.7	14.4
50 × 100	5.00	83.3	41.7	4.17	1.04	28.9	14.4
50 × 125	6.25	130	52.1	8.14	1.30	36.1	14.4
50 × 150	7.50	188	62.5	14.1	1.56	43.3	14.4
50 × 175	8.75	255	72.9	22.3	1.82	50.5	14.4
50 × 200	10.0	333	83.3	33.3	2.08	57.7	14.4
50 × 225	11.3	422	93.8	47.5	2.34	65.0	14.4
50 × 250	12.5	521	104	65.1	2.60	72.2	14.4
50 × 300	15.0	750	125	113	3.13	86.6	14.4
63 × 100	6.30	105	66.2	5.25	2.08	28.9	18.2
63 × 125	7.88	164	82.7	10.3	2.60	36.1	18.2
63 × 150	9.45	236	99.2	17.7	3.13	43.3	18.2
63 × 175	11.0	322	116	28.1	3.65	50.5	18.2
63 × 200	12.6	420	132	42.0	4.17	57.7	18.2
63 × 225	14.2	532	149	59.8	4.69	65.0	18.2

Table 2.4   Geometrical properties of sawn softwoods (BS 5268 Part 2 1988 Table 98) (continued)

Basic size* (mm)	Area (10^3 mm^2)	Section modulus		Second moment of area		Radius of gyration	
		About x–x (10^3 mm^3)	About y–y (10^3 mm^3)	About x–x (10^6 mm^4)	About y–y (10^6 mm^4)	About x–x (mm)	About y–y (mm)
75 × 100	7.50	125	93.8	6.25	3.52	28.9	21.7
75 × 125	9.38	195	117	12.2	4.39	36.1	21.7
75 × 150	11.3	281	141	21.1	5.27	43.3	21.7
75 × 175	13.1	383	164	33.5	6.15	50.5	21.7
75 × 200	15.0	500	188	50.0	7.03	57.7	21.7
75 × 225	16.9	633	211	71.2	7.91	65.0	21.7
75 × 250	18.8	781	234	97.7	8.79	72.2	21.7
75 × 300	22.5	1130	281	169	10.5	86.6	21.7
100 × 100	10.0	167	167	8.33	8.33	28.9	28.9
100 × 150	15.0	375	250	28.1	12.5	43.3	28.9
100 × 200	20.0	667	333	66.7	16.7	57.7	28.9
100 × 250	25.0	1040	417	130	20.8	72.2	28.9
100 × 300	30.0	1500	500	225	25.0	86.6	28.9
150 × 150	22.5	563	563	42.2	42.2	43.3	43.3
150 × 200	30.0	1000	750	100	56.3	57.7	43.3
150 × 300	45.0	2250	1130	338	84.4	86.6	43.3
200 × 200	40.0	1330	1330	133	133	57.7	57.7
250 × 250	62.5	2600	2600	326	326	72.2	72.2
300 × 300	90.0	4500	4500	675	675	86.6	86.6

* Basic size measured at 20 per cent moisture content.

Table 2.5   Modification factor K_1 by which the geometrical properties of timber for the dry exposure condition should be multiplied to obtain values for the wet exposure condition (BS 5268 Part 2 1988 Table 2)

Geometrical property	Value of K_1
Thickness, width, radius of gyration	1.02
Cross-sectional area	1.04
First moment of area, section modulus	1.06
Second moment of area	1.08

value is permitted by multiplying it by a load sharing modification factor K_8 of 1.1.

2.11   Types of member

There are certain design considerations which apply specifically to either flexural members (such as beams) or compression members (such as posts). These will be explained in greater detail in the following sections.

Table 2.6   Modification factor K_3 for duration of loading (BS 5268 Part 2 1988 Table 17)

Duration of loading	Value of K_3
Long term (e.g. dead + permanent imposed*)	1.00
Medium term (e.g. dead + snow, dead + temporary imposed)	1.25
Short term (e.g. dead + imposed + wind,† dead + imposed + snow + wind†)	1.50
Very short term (e.g. dead + imposed + wind‡)	1.75

* For uniformly distributed imposed floor loads $K_3 = 1$ except for type 2 and type 3 buildings in Table 5 of BS 6399 Part 1 1984 (here Table 1.5) where, for corridors, hallways, landings and stairways only, K_3 may be assumed to be 1.5.

† For wind, short term category applies to class C (15 s gust) as defined in CP 3 Chapter V Part 2.

‡ For wind, very short term category applies to classes A and B (3 s or 5 s gust) as defined in CP 3 Chapter V Part 2.

2.12   Flexural members

Flexural members are those subjected to bending, for example beams, rafters, joists and purlins. The main design considerations for which flexural members should be examined are

(a) Bending (including lateral buckling)

(b) Deflection

(c) Shear

(d) Bearing.

Generally, bending is the critical condition for medium span beams, deflection for long span beams, and shear for heavily loaded short span beams or at notched ends.

Let us now examine how we consider each of these in design.

2.12.1   Bending (including lateral buckling)

For designs based on permissible stress philosophy, bending is checked by applying the basic theory of bending principles. In relation to timber design this must also take into account the relevant modification factors for exposure, load duration, load sharing and so on.

From the theory of bending we know that $M = fZ$, where $Z = bd^2/6$ for rectangular sections. Knowing the applied loads, the maximum bending moment M may be calculated. Hence the required section modulus Z about the x–x axis may be obtained:

$$Z_{xx} \text{ required} = \frac{M}{f}$$

The symbol f denotes the permissible stress value of the material, which for timber flexural members is the grade bending stress $\sigma_{m,g,par}$ modified by any relevant K factors. These are the previously mentioned K_2 exposure factor (when applicable), the K_3 load duration factor and the K_8 load sharing factor (if applicable), together with a depth factor K_7.

The depth factor K_7 is necessary because the grade bending stresses given in Table 2.2 only apply to timber sections with a depth h of 300 mm. For depths of 72 mm or less the grade bending stress should be multiplied by a K_7 factor of 1.17. For depths h greater than 72 mm and less than 300 mm the grade bending stress has to be multiplied by a depth factor K_7 obtained from the following expression:

$$K_7 = \left(\frac{300}{h}\right)^{0.11}$$

Values of K_7 computed from this expression for the range of sawn, planed and regularized timber sections generally available are given in Table 2.7.

Table 2.7 Modification factor K_7 for depth

h	$K_7 - (300/h)^{0.11}$
72	1.170
75	1.165
97	1.132
100	1.128
120	1.106
122	1.104
125	1.101
145	1.083
147	1.082
150	1.079
169	1.065
170	1.064
175	1.061
194	1.049
195	1.049
200	1.046
219	1.035
220	1.034
225	1.032
244	1.023
245	1.023
250	1.020
294	1.002
295	1.002
300	1.000

Thus the expression for calculating the section modulus Z_{xx} for timber members, incorporating all the K factors, is as follows:

$$Z_{xx} \text{ required} = \frac{M}{\sigma_{m,g,par} K_2 K_3 K_7 K_8}$$

A suitable section size having a Z_{xx} greater than that required may then be chosen from one of BS 5268 Tables 98, 99 or 100. The chosen section should then be checked for deflection and shear.

It is also necessary to ensure that whilst the member is bending vertically, lateral buckling failure is not induced. To avoid such failure, the recommended depth to breadth ratio values given in BS 5268 Table 19 should be complied with. This table is reproduced here as Table 2.8.

Table 2.8   Maximum depth to breadth ratios (solid and laminated members) (BS 5268 Part 2 1988 Table 19)

Degree of lateral support	Maximum depth to breadth ratio
No lateral support	2
Ends held in position	3
Ends held in position and member held in line as by purlins or tie rods at centres not more than 30 times breadth of the member	4
Ends held in position and compression edge held in line, as by direct connection of sheathing, deck or joists	5
Ends held in position and compression edge held in line, as by direct connection of sheathing, deck or joists, together with adequate bridging or blocking spaced at intervals not exceeding 6 times the depth	6
Ends held in position and both edges held firmly in line	7

2.12.2   Deflection

To avoid damage to finishes, ceilings, partitions and so on, the deflection of timber flexural members when fully loaded should be limited to 0.003 of the span. In addition, for longer span domestic floors (over 4.67 m) the maximum deflection should not exceed 14 mm. That is, the permissible deflection δ_p is as follows:

Generally.

$$\delta_p = 0.003 \times \text{span}$$

For long span domestic floors:

$$\delta_p = 0.003 \times \text{span} \quad \text{but} \ngtr 14 \text{mm}$$

For a flexural member to be adequate in deflection the summation of the actual deflection due to bending δ_m and that due to shear δ_v must not be greater than the permissible value δ_p:

$$\delta_m + \delta_v \ngtr \delta_p$$

The actual bending deflection δ_m is calculated using the formula relevant to the applied loading:

For a UDL:

$$\delta_m = \frac{5}{384} \frac{WL^3}{EI}$$

For a central point load:

$$\delta_m = \frac{1}{48} \frac{WL^3}{EI}$$

The E value to be adopted is E_{min} for isolated members and E_{mean} for load sharing members.

The actual deflection δ_v produced by shear on rectangular (and square) cross-section members is calculated using the following expression:

$$\delta_v = \frac{SM}{AG}$$

where

M mid-span bending moment
S member shape factor, which is 1.2 for rectangular sections
A section area
G modulus of rigidity or shear modulus, taken as $E/16$
E appropriate E value: E_{min} for isolated members, E_{mean} for load sharing members

The expression can be rewritten to include the values for S and G as follows:

$$\delta_v = \frac{SM}{AG} = \frac{1.2M}{AE/16} = \frac{19.2M}{AE}$$

The total actual deflection δ_a resulting from bending and shear will therefore be the summation of the relevant formulae:

$$\delta_a = \delta_m + \delta_v$$

For a UDL:

$$\delta_a = \frac{5}{384} \frac{WL^3}{EI} + \frac{19.2M}{AE}$$

For a point load:

$$\delta_a = \frac{1}{48} \frac{WL^3}{EI} + \frac{19.2M}{AE}$$

2.12.3 Shear

The critical position for shear on a normally loaded flexural member is at the support where the maximum reaction occurs. The shear stress occurring at that position is calculated and compared with the permissible value.

For rectangular timber flexural members the maximum applied shear stress parallel to the grain, r_a, occurs at the NA and is calculated from the following expression:

$$r_a = \frac{3}{2}\frac{F_v}{A}$$

where F_v is the maximum vertical shear (usually maximum reaction) and A is the cross-sectional area (bh).

The applied shear stress must be less than the permissible shear stress parallel to the grain, r_{adm}. This is obtained by multiplying the grade shear stress parallel to the grain, r_g (Table 2.2), by the K_3 load duration and K_8 load sharing factors as appropriate. Thus

$$r_a \leqslant r_{adm}$$
$$r_a \leqslant r_g K_3 K_8$$

Notches occurring at the ends of flexural members affect their shear capacity. To allow for this, the permissible shear stress parallel to the grain is further reduced by a modification factor K_5. Thus

$$r_a = \frac{3}{2}\frac{F_v}{A} \leqslant r_{adm} K_5$$

The factor K_5 is obtained in one of two ways depending on whether the notch occurs in the top (Figure 2.1) or the bottom (Figure 2.2) of the member:

Top edge notches: $K_5 = \begin{cases} \dfrac{h(h_e - a) + ah_e}{h_e^2} & \text{when} \quad a \leqslant h_e \\ 1.0 & \text{when} \quad a > h_e \end{cases}$

Bottom edge notches: $K_5 = \dfrac{h_e}{h}$

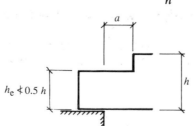

Figure 2.1 *Top edge notch*   **Figure 2.2** *Bottom edge notch*

The effective depth h_e of timber remaining should never be less than half the member depth, that is $0.5h$.

It should be appreciated that the area used in the expression for calculating r_a at a notched end is the net area after notching, that is $A = bh_e$.

2.12.4   Bearing

The compression stress perpendicular to the grain, or the bearing stress, which is developed at points of support should also be checked. Where the length of bearing at a support is not less than 75 mm the bearing stress will not usually be critical. If flexural members are supported on narrow beams or ledgers the bearing stress could influence the member size.

The applied compression stress perpendicular to the grain, $\sigma_{c,a,perp}$, may be calculated from the following expression:

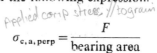

$$\sigma_{c,a,perp} = \frac{F}{\text{bearing area}}$$

where F is the bearing force, usually maximum reaction, and the bearing area is the bearing length times the breadth of the section.

This stress must be less than the permissible compression stress perpendicular to the grain, $\sigma_{c,adm,perp}$. This is obtained by multiplying the grade compression stress perpendicular to the grain, $\sigma_{c,g,perp}$ (Table 2.2), by the K_3 load duration and K_8 load sharing factors as appropriate. Therefore

$$\sigma_{c,a,perp} \leqslant \sigma_{c,adm,perp}$$
$$\sigma_{c,a,perp} \leqslant \sigma_{c,g,perp} K_3 K_8$$

Two values for the grade compression stress perpendicular to the grain are given for each strength class in Table 2.2. The higher value may be used when the specification will prohibit wane from occurring at bearing areas; otherwise the lower value must be adopted.

Reference should be made to BS 5268 for the modification factor K_4, which should be used for the special case of bearing less than 150 mm long located 75 mm or more from the end of a flexural member.

2.12.5   Design summary for timber flexural members

The design procedure for timber flexural members such as beams, joists and rafters may be summarized as follows.

Bending

Check using the theory of bending principles, taking into account modification factors for load duration K_3, load sharing K_8 (if applicable), depth

K_7 and so on. First, calculate the bending moment M. Then

$$\text{Approximate } Z_{xx} \text{ required} = \frac{M}{\sigma_{m,g,par} K_3 K_8}$$

Choose a suitable timber section from tables, and recheck the Z_{xx} required with a K_7 factor for depth included:

$$\text{Final } Z_{xx} \text{ required} = \frac{M}{\sigma_{m,g,par} K_3 K_7 K_8}$$

The risk of lateral buckling failure should also be checked at this stage by ensuring that the depth to breadth ratio is less than the relative maximum value given in Table 2.8.

Deflection

Compare the permissible deflection $\delta_p = 0.003 \times \text{span}$ with the actual deflection $\delta_a = \delta_m + \delta_v$:

For a UDL:
$$\delta_a = \frac{5}{384} \frac{WL^3}{EI} + \frac{19.2M}{AE}$$

For a central point load:
$$\delta_a = \frac{1}{48} \frac{WL^3}{EI} + \frac{19.2M}{AE}$$

Use E_{mean} for load sharing members and E_{min} for isolated members.

Shear

Compare the applied shear stress

$$r_a = \frac{3}{2} \frac{F_v}{A}$$

with the permissible shear stress

$$r_{adm} = r_g K_3 K_8$$

(where K_8 is used if applicable). Then if the member is notched, check the shear with a notch.

Bearing

Compare the applied bearing stress

$$\sigma_{c,a,perp} = \frac{F}{\text{bearing area}}$$

with the permissible bearing stress

$$\sigma_{c,adm,perp} = \sigma_{c,g,perp} K_3 K_8$$

(where K_8 is used if applicable).

Note that throughout this procedure the wet exposure modification factors should be applied where necessary if the member is to be used externally.

Let us now look at some examples on the design of timber flexural members.

Example 2.1

A flat roof spanning 4.25 m is to be designed using timber joists at 600 m centres. The loads from the proposed roof construction are as follows:

Asphalt 20 mm thick	$0.45 \, kN/m^2$
Pre-screeded unreinforced woodwool	$0.30 \, kN/m^2$
Timber firrings	$0.01 \, kN/m^2$
Ceiling	$0.15 \, kN/m^2$

The imposed roof load due to snow may be taken as $0.75 \, kN/m^2$ and the load due to the weight of the joists as $0.1 \, kN/m^2$. The latter may be checked after the joists have been designed by taking the weight of timber as $540 \, kg/m^3$.

Determine the size of suitable SC3 whitewood joists, checking shear and deflection. In addition, check the joists if a 75 mm deep notch were to be provided to the bottom edge at the bearings.

Loading

Before proceeding with the design routine for flexural members, the load carried by a joist must be computed in the manner described in Chapter 1:

Dead load:	asphalt	0.45
	woodwool	0.30
	firrings	0.01
	ceiling	0.15
	joists SW	$\underline{0.10}$
		$1.01 \, kN/m^2$

Imposed load: $0.75 \, kN/m^2$

Combined load:	dead	1.01
	imposed	$\underline{0.75}$
		$1.76 \, kN/m^2$

From this the uniformly distributed load supported by a single joist can be calculated on the basis that they span 4.25 m and are spaced at 600 m centres.

UDL per joist $= 1.76 \times 4.25 \times 0.6 = 4.49 \, kN$,   say $4.5 \, kN$

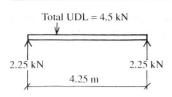

Total UDL = 4.5 kN

2.25 kN   2.25 kN

4.25 m

Figure 2.3   *Loaded joist*

This enables us to visualize the load condition for a single roof joist element in Figure 2.3.

The design procedure for timber flexural members can now be followed.

Bending

Maximum bending moment $M = \dfrac{WL}{8} = \dfrac{4.5 \times 4.25}{8} = 2.39 \, \text{kN m}$

$$= 2.39 \times 10^6 \, \text{N mm}$$

Grade bending stress parallel to grain (from Table 2.2 for SC3 whitewood timber) $\sigma_{m,g,par} = 5.3 \, \text{N/mm}^2$

K_3 load duration factor, medium term (from Table 2.6) $= 1.25$

K_8 load sharing factor $= 1.1$

K_7 depth factor is unknown at this stage

Approximate Z_{xx} required (ignoring K_7)

$$= \frac{M}{\sigma_{m,g,par} K_3 K_8} = \frac{2.39 \times 10^6}{5.3 \times 1.25 \times 1.1} = 327\,959 \, \text{mm}^3 = 327.96 \times 10^3 \, \text{mm}^3$$

Comparing this Z_{xx} required with the properties for sawn joists given in Table 2.4, a 50 mm × 200 mm joist has a Z_{xx} of $333 \times 10^3 \, \text{mm}^3$, which is greater than that required. Therefore the section would be adequate in bending.

If the K_7 depth factor were to be included in the calculations now that a timber size has been determined, it would increase the adequacy of the chosen section since the approximate Z_{xx} required is divided by this factor. From Table 2.7 for a 200 mm joist, $K_7 = 1.046$. Therefore

$$\text{Final } Z_{xx} \text{ required} = \frac{\text{approximate } Z_{xx} \text{ required}}{K_7}$$

$$= \frac{327.96 \times 10^6}{1.046} = 313.54 \times 10^3 \, \text{mm}^3$$

If it were considered necessary, a comparison could be made between the applied bending stress developed in the timber and the permissible bending stress. Such a comparison is not essential since the section has already been shown to be adequate in bending, but it will be included here to illustrate the calculation involved.

Applied bending stress parallel to grain:

$$\sigma_{m,a,par} = \frac{M}{Z_{xx}} = \frac{2.39 \times 10^6}{333 \times 10^3} = 7.18 \, \text{N/mm}^2$$

Permissible bending stress parallel to grain:

$$\sigma_{m,adm,par} = \sigma_{m,g,par} K_3 K_7 K_8 = 5.3 \times 1.25 \times 1.046 \times 1.1 = 7.62 \, \text{N/mm}^2$$

The risk of lateral buckling failure should also be considered:

Depth to breadth ratio $h/d = 200/50 = 4$

By comparison with Table 2.8, the maximum ratio for this form of construction would be 5. Therefore the section satisfies lateral buckling requirements.

Deflection

Permissible deflection $\delta_p = 0.003 \times \text{span} = 0.003 \times 4250 = 12.75\,\text{mm}$

Actual deflection $\delta_a = \delta_m + \delta_v = \dfrac{5}{384}\dfrac{WL^3}{EI} + \dfrac{19.2M}{AE}$

$$= \frac{5}{384} \times \frac{4.5 \times 10^3 \times 4250^3}{8800 \times 33.3 \times 10^6} + \frac{19.2 \times 2.39 \times 10^6}{10 \times 10^3 \times 8800}$$

$$= 15.35 + 0.52 = 15.87\,\text{mm} > 12.75\,\text{mm}$$

Therefore the 50 mm × 200 mm joist is not adequate in deflection.

As a guide to choosing a suitable section for deflection, the I_{xx} required to satisfy bending deflection alone (ignoring shear deflection) can be calculated in relation to the permissible deflection. We have

$$\delta_m - \frac{5}{384}\frac{WL^3}{EI}$$

Thus for δ_m only, rewriting this equation and using $\delta_p = 12.75\,\text{mm}$,

$$I_{xx} \text{ required} - \frac{5}{384}\frac{WL^3}{E\delta_p}$$

$$= \frac{5}{384} \times \frac{4.5 \times 10^3 \times 4250^3}{8800 \times 12.75} = 40.09 \times 10^6\,\text{mm}^4$$

Comparing this with the I_{xx} properties for sawn joists given in Table 2.4 shows that the previous 50 mm × 200 mm joist has an I_{xx} of $33.3 \times 10^6\,\text{mm}^4$ and would not suffice. By inspection of the table a more reasonable section to check would be 50 mm × 225 mm with an I_{xx} of $47.5 \times 10^6\,\text{mm}^4$. Hence

$$\delta_a = \frac{5}{384} \times \frac{4.5 \times 10^3 \times 4250^3}{8800 \times 47.5 \times 10^6} + \frac{19.2 \times 2.39 \times 10^6}{11.3 \times 10^3 \times 8800}$$

$$= 10.76 + 0.46 = 11.22\,\text{mm} < 12.75\,\text{mm}$$

Therefore the 50 mm × 225 mm joist is adequate in deflection, and will also be adequate in bending.

It would perhaps be simpler in future to adopt this approach for determining a trial section at the beginning of the deflection check.

Shear unnotched

Maximum shear $F_v = \text{reaction} = \dfrac{\text{UDL}}{2} = \dfrac{4.5}{2} - 2.25\,\text{kN} = 2.25 \times 10^3\,\text{N}$

Grade shear stress parallel to grain (from Table 2.2) $r_g = 0.67\,\text{N/mm}^2$

Permissible shear stress $r_{adm} = r_g K_3 K_8 = 0.67 \times 1.25 \times 1.1 = 0.92\,\text{N/mm}^2$

Applied shear stress $r_a = \dfrac{3}{2}\dfrac{F_v}{A} = \dfrac{3}{2} \times \dfrac{2.25 \times 10^3}{11.3 \times 10^3} = 0.3\,\text{N/mm}^2 < 0.92\,\text{N/mm}^2$

Thus the section is adequate in shear unnotched.

Shear notched

Check the section with a 75 mm deep bottom edge notch at the bearing as illustrated in Figure 2.4. The permissible shear stress for a member notched in this manner must be multiplied by a further modification factor K_5. For bottom edge notches,

$$K_5 = \frac{h_e}{h} = \frac{150}{225} = 0.67$$

Hence

$$r_{\text{adm}} = r_g K_3 K_8 K_5 = 0.92 \times 0.67 = 0.62 \, \text{N/mm}^2$$

$$r_a = \frac{3}{2} \frac{F_v}{bh_e} = \frac{3}{2} \times \frac{2.25 \times 10^3}{50 \times 150} = 0.45 \, \text{N/mm}^3 < 0.62 \, \text{N/mm}^2$$

Therefore the section is also adequate in shear when notched as shown in Figure 2.4.

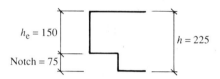

$h_e = 150$

Notch = 75

$h = 225$

Figure 2.4   *Notched joist*

Bearing

Maximum bearing force F = reaction = $2.25 \times 10^3 \, \text{N}$

Assuming that the roof joists span on to a 100 mm wide wall plate, the bearing length will be 100 mm and hence the bearing area will be this length multiplied by the section breadth.

Applied bearing stress $\sigma_{c,a,\text{perp}} = \dfrac{F}{\text{bearing area}} = \dfrac{2.25 \times 10^3}{100 \times 50} = 0.45 \, \text{N/mm}^2$

The grade bearing stress is the compression stress perpendicular to the grain from Table 2.2. This will be taken as the higher of the two values on the basis that wane will be specifically excluded at bearing positions. Therefore

Grade bearing stress $\sigma_{c,g,\text{perp}} = 2.2 \, \text{N/mm}^2$

Permissible bearing stress $\sigma_{c,\text{adm},\text{perp}} = \sigma_{c,g,\text{perp}} K_3 K_8$
$$= 2.2 \times 1.25 \times 1.1$$
$$= 3.03 \, \text{N/mm}^2 > 0.45 \, \text{N/mm}^2$$

The section is adequate in bearing

Joist self-weight

Finally, the load that was assumed for the joists can be verified now that the size is known and given that the timber weighs 540 kg/m³:

$$\text{SW of joists at 600 centres} = \frac{540}{100} \times 0.225 \times 0.05 \times \frac{1000}{600} = 0.1 \, \text{kN/m}^2$$

$$\text{SW assumed} = 0.1 \, \text{kN/m}^2$$

Conclusion

Use 50 mm × 225 mm SC3 whitewood sawn joists.

Example 2.2

Design the timber floor for a dwelling if it comprises tongued and grooved (T&G) boards carried by 3.6 m span joists at 600 mm centres. The load imposed by the dead weight of the boards is 0.1 kN/m², by the joists 0.12 kN/m² and by a plaster ceiling on the underside 0.18 kN/m². The floor is subjected to a domestic imposed load of 1.5 kN/m².

Use home grown Douglas fir M50/SS timber.

Loading

Dead load: boards 0.10
          joists 0.12
          ceiling 0.18
                  0.4 kN/m²

Imposed load: 1.5 kN/m²

Combined load: dead   0.4
              imposed 1.5
                      1.9 kN/m²

Guidance on the specification of T&G softwood flooring is given in BS 1297. The thickness of T&G floor boards for domestic situations may be obtained directly from the Building Regulations. The board thickness recommended for joists spaced at 600 mm is 19 mm.

$$\text{UDL per joist} = 1.9 \times 3.6 \times 0.6 = 4.1 \, \text{kN} = 4.1 \times 10^3 \, \text{N}$$

Bending

$$M = \frac{WL}{8} = \frac{4.1 \times 3.6}{8} = 1.85 \, \text{kN m} = 1.85 \times 10^6 \, \text{N mm}$$

$\sigma_{m,g,par}$ for M50/SS = 5.3 N/mm²

K_3 (long term) = 1.0; $K_8 = 1.1$; K_7 is unknown

Approximate Z_{xx} required

$$= \frac{M}{\delta_{m,g,par} K_3 K_8} = \frac{1.85 \times 10^6}{5.3 \times 1.0 \times 1.1} = 317\,324 \, \text{mm}^3 = 317 \times 10^3 \, \text{mm}^3$$

By reference to Table 2.8, the maximum depth to breadth ratio needed to ensure lateral stability is 5.

From Table 2.4, a 50 mm × 200 mm joist has $Z_{xx} = 333 \times 10^3 \, \text{mm}^3$. Check with $K_7 = 1.046$:

$$\text{Final } Z_{xx} \text{ required} = \frac{317 \times 10^3}{1.046} = 303 \times 10^3 \, \text{mm}^3$$

Deflection

Permissible $\delta_p = 0.003 \times \text{span} = 0.003 \times 3600 = 10.8\,\text{mm}$

$$\text{Actual } \delta_a = \delta_m + \delta_v = \frac{5}{384}\frac{WL^3}{EI} \times \frac{19.2M}{AE}$$

$$= \frac{5}{384} \times \frac{4.1 \times 10^3 \times 3600^3}{8800 \times 33.3 \times 10^6} + \frac{19.2 \times 1.85 \times 10^6}{10 \times 10^3 \times 8800}$$

$$= 8.5 + 0.4 = 8.9\,\text{mm} < 10.8\,\text{mm}$$

Thus the 50 mm × 200 mm joist is adequate in deflection.

Shear unnotched

Maximum shear $F_v = \dfrac{UDL}{2} = \dfrac{4.1}{2} = 2.05\,\text{kN} = 2.05 \times 10^3\,\text{N}$

$r_g = 0.67\,\text{N/mm}^2$

$r_{adm} = r_g K_3 K_8 = 0.67 \times 1 \times 1.1 = 0.737\,\text{N/mm}^2$

$r_a = \dfrac{3}{2}\dfrac{F_v}{A} = \dfrac{3}{2} \times \dfrac{2.05 \times 10^3}{10 \times 10^3} = 0.308\,\text{N/mm}^2 < 0.737\,\text{N/mm}^2$

Thus the 50 mm × 200 mm joist is adequate in shear unnotched.

Bearing

$F = 2.05 \times 10^3\,\text{N}$

Assume that the joists are supported on 100 mm blockwork; hence the bearing length will be 100 mm.

$\sigma_{c,a,perp} = \dfrac{F}{\text{bearing area}} = \dfrac{2.05 \times 10^3}{100 \times 50} = 0.41\,\text{N/mm}^2$

$\sigma_{c,g,perp} = 2.2\,\text{N/mm}^2,$ wane prohibited

$\sigma_{c,adm,perp} = \sigma_{c,g,perp} K_3 K_8 = 2.2 \times 1 \times 1.1 = 2.42\,\text{N/mm}^2 > 0.41\,\text{N/mm}^2$

The section is adequate in bearing.

Conclusion

Use 50 mm × 200 mm M50/SS home grown Douglas fir joists.

Example 2.3

Timber roof purlins spanning 2.65 m support a total UDL, inclusive of their own weight, of 9 kN. Using GS grade redwood, what size of member is required?

Loading

Total UDL = 9 kN

Bending

$$M = \frac{WL}{8} = \frac{9 \times 2.65}{8} = 2.98\,\text{kN m} = 2.98 \times 10^6\,\text{N mm}$$

$\sigma_{m,g,par} = 5.3\,N/mm^2$

K_3 (medium term) $= 1.25$

Purlins will be spaced at centres greater than 600 mm and therefore are not load sharing; hence K_8 factor does not apply.

Approximate Z_{xx} required

$$= \frac{M}{\sigma_{m,g,par}K_3} = \frac{2.98 \times 10^6}{5.3 \times 1.25} = 449\,811\,mm^3 = 450 \times 10^3\,mm^3$$

Maximum depth to breadth ratio to avoid lateral buckling is 5.

From Table 2.4:

For 75 mm × 200 mm sawn joists: $Z_{xx} = 500 \times 10^3\,mm^3$
For 63 mm × 225 mm sawn joists: $Z_{xx} = 532 \times 10^3\,mm^3$
For 75 mm × 225 mm sawn joists: $Z_{xx} = 633 \times 10^3\,mm^3$

Check with $K_7 = 1.032$:

Final Z_{xx} required $= \dfrac{450 \times 10^3}{1.032} = 436 \times 10^3\,mm^3$

Deflection
Permissible $\delta_p = 0.003 \times span = 0.003 \times 2650 = 7\,95\,mm$

Since purlins are not load sharing, E_{min} must be used when calculating the actual deflection.

Actual $\delta_a = \delta_m + \delta_v = \dfrac{5}{384}\dfrac{WL^3}{EI} + \dfrac{19.2M}{AE}$

For 75 × 200: $\delta_a = \dfrac{5}{384} \times \dfrac{9 \times 10^3 \times 2650^3}{5800 \times 50 \times 10^6} + \dfrac{19.2 \times 2.98 \times 10^6}{15 \times 10^3 \times 5800}$

$\qquad\qquad = 7.5 + 0.66 = 8.16\,mm > 7.95\,mm$

For 63 × 225: $\delta_a = \dfrac{5}{384} \times \dfrac{9 \times 10^3 \times 2650^3}{5800 \times 59.8 \times 10^6} + \dfrac{19.2 \times 2.98 \times 10^6}{14.2 \times 10^3 \times 5800}$

$\qquad\qquad = 6.29 + 0.69 = 6.98\,mm < 7.95\,mm$

For 75 × 225: $\delta_a = \dfrac{5}{384} \times \dfrac{9 \times 10^3 \times 2650^3}{5800 \times 71.2 \times 10^6} + \dfrac{19.2 \times 2.98 \times 10^6}{16.9 \times 10^3 \times 5800}$

$\qquad\qquad = 5.28 + 0.58 = 5.86\,mm < 7.95\,mm$

Thus both 63 mm × 225 mm and 75 mm × 225 mm joists are adequate.

Shear unnotched
Maximum shear $F_v = 4.5\,kN = 4.5 \times 10^3\,N$

$r_g = 0.67\,N/mm^2$

$r_{adm} = r_g K_3 = 0.67 \times 1.25 = 0.84\,N/mm^2$

For 63×225: $r_a = \dfrac{3}{2} \times \dfrac{4.5 \times 10^3}{14.2 \times 10^3} = 0.48 \,\text{N/mm}^2 < r_{adm}$

For 75×225: $r_a = \dfrac{3}{2} \times \dfrac{4.5 \times 10^3}{16.9 \times 10^3} = 0.4 \,\text{N/mm}^2 < r_{adm}$

Both sections are therefore adequate.

Bearing

$F = 4.5 \times 10^3 \,\text{N}$

Assume that the purlins will be supported on 100 mm blockwork and check the narrower choice of section:

$$\sigma_{c,a,perp} = \frac{F}{\text{bearing area}} = \frac{4.5 \times 10^3}{100 \times 63} = 0.71 \,\text{N/mm}^2$$

$\sigma_{c,g,perp} = 2.2 \,\text{N/mm}^2$, wane prohibited

$\sigma_{c,adm,perp} = \sigma_{c,g,perp} K_3 = 2.2 \times 1.25 = 2.75 \,\text{N/mm}^2 > 0.71 \,\text{N/mm}^2$

Both sections are adequate.

Conclusion

Use 63 mm $\times$ 225 mm or 75 mm $\times$ 225 mm SC3 redwood sawn purlins. The final choice may be determined by availability.

Example 2.4

Timber roof beams spaced on a grid of 1200 mm are required to span 7.2 m, supporting a total dead plus imposed load of $1.5 \,\text{kN/m}^2$. What size of solid timber joist, having a grade bending stress of $5.3 \,\text{N/mm}^2$ and a minimum E of 5800 N/mm^2, would be required?

Total UDL $= 1.5 \times 7.2 \times 1.2 = 12.96 \,\text{kN}$

$$M = \frac{WL}{8} = \frac{12.96 \times 7.2}{8} = 11.66 \,\text{kN m} = 11.66 \times 10^6 \,\text{N mm}$$

K_3 (medium term) $= 1.25$; K_7 is unknown; K_8 load sharing factor is not applicable.

Approximate Z_{xx} required $= \dfrac{11.66 \times 10^6}{5.3 \times 1.25} = 1\,760\,000 \,\text{mm}^3 = 1760 \times 10^3 \,\text{mm}^3$

$\delta_p = 0.003 \times \text{span} = 0.003 \times 7200 = 21.6 \,\text{mm}$

Approximate I_{xx} required for δ_m

$$= \frac{5}{384} \frac{WL^3}{E\delta_p} = \frac{5}{384} \times \frac{12.96 \times 10^3 \times 7200^3}{5800 \times 21.6} = 502\,758\,620 \,\text{mm}^4 = 503 \times 10^6 \,\text{mm}^4$$

By reference to Table 2.4, only a 300 mm $\times$ 300 mm solid timber section would appear to be adequate. This however would not normally be considered to be a practical choice for a beam. The alternative is to use a stronger hardwood section, or one of the many proprietary timber beams available.

The design of hardwood flexural members follows exactly the same procedure as that for softwoods, but of course the higher grade stresses given in BS 5268 for hardwood species are used. Proprietary timber beams will be discussed in more detail in the following section.

2.13 Proprietary timber beams

These are generally used in situations where the structural capacity of solid softwood timber sections is exceeded, perhaps due to long spans or wide spacing. They are available from a number of timber suppliers and consequently their exact shape or form will depend on the individual manufacturer. Three of the types generally offered are illustrated in Figure 2.5: (a) glued laminated or glulam, (b) plywood box beams and (c) ply web beams.

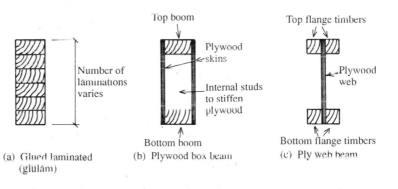

(a) Glued laminated (glulam)   (b) Plywood box beam   (c) Ply web beam

Figure 2.5 *Proprietary timber beams*

Such beams are designed in accordance with BS 5268, but the supplier normally produces design literature in the form of safe load tables. These give the safe span for each beam profile in relation to the applied loading. The user therefore only needs to calculate the design load in the usual manner and then choose a suitable beam section from the tables.

If desired the structural engineer could of course design individual beams in accordance with BS 5268 to suit his own requirements. The production of the beams would then be undertaken by a specialist timber manufacturer.

2.14 Compression members: posts

Compression members include posts or columns, vertical wall studs, and the struts in trusses and girders. The design of single isolated posts and load bearing stud walls will be considered in this manual, beginning in this section with posts.

It is important when selecting suitable pieces of timber for use as columns that particular attention is paid to straightness. The amount of bow permitted by most stress grading rules is not usually acceptable for the selection of column material. The amount of bow acceptable for column members should be limited to 1/300 of the length.

Timber posts may be subject to direct compression alone, where the loading is applied axially, or to a combination of compression loading and bending due to the load being applied eccentrically to the member axes. A timber post may also have to be designed to resist lateral bending resulting from wind action. However, the effects of wind loading on individual structural elements will not be considered in this manual.

The structural adequacy of an axially loaded post is determined by comparing the applied compression stress parallel to the grain with the permissible compression stress parallel to the grain.

2.14.1   Applied compression stress

The applied stress parallel to the grain is obtained by dividing the applied load by the cross-sectional area of the timber section:

$$\sigma_{c,a,par} = \frac{\text{applied load}}{\text{section area}} = \frac{F}{A}$$

The section area is the net area after deducting any open holes or notches. No deduction is necessary for holes containing bolts.

For the section to be adequate, the applied stress must be less than the permissible stress:

$$\sigma_{c,a,par} < \sigma_{c,adm,par}$$

2.14.2   Permissible compression stress

The permissible stress $\sigma_{c,adm,par}$ is obtained by modifying the grade compression stress parallel to the grain, $\sigma_{c,g,par}$ (Table 2.2), by any of the previously mentioned K factors that may be applicable, that is

K_1 wet exposure geometrical property modification factor
K_2 wet exposure stress modification factor
K_3 load duration modification factor

Timber posts, as opposed to wall studs, are not normally part of a load sharing system as defined by BS 5268 and therefore the load sharing modification factor K_8 does not apply.

2.14.3   Slenderness of posts

To avoid lateral buckling failure a further modification factor must also be applied in post calculations when the slenderness ratio is equal to 5 or more. This is obtained from BS 5268 Table 22, reproduced here as Table 2.9. It is dependent on the slenderness ratio and on the ratio of the modulus of elasticity to the compression stress (E/σ).

Table 2.9 Modification factor K_{12} for compression members (BS 5268 Part 2 1988 Table 22)

$E/\sigma_{c,\|}$	\<5	5	10	20	30	40	50	60	70	80	90	100	120	140	160	180	200	220	240	250
	\<1.4	1.4	2.9	5.8	8.7	11.6	14.5	17.3	20.2	23.1	26.0	28.9	34.7	40.5	46.2	52.0	57.8	63.6	69.4	72.3
400	1.000	0.975	0.951	0.896	0.827	0.735	0.621	0.506	0.408	0.330	0.271	0.225	0.162	0.121	0.094	0.075	0.061	0.051	0.043	0.040
500	1.000	0.975	0.951	0.899	0.837	0.759	0.664	0.562	0.466	0.385	0.320	0.269	0.195	0.148	0.115	0.092	0.076	0.063	0.053	0.049
600	1.000	0.975	0.951	0.901	0.843	0.774	0.692	0.601	0.511	0.430	0.363	0.307	0.226	0.172	0.135	0.109	0.089	0.074	0.063	0.058
700	1.000	0.975	0.951	0.902	0.848	0.784	0.711	0.629	0.545	0.467	0.399	0.341	0.254	0.195	0.154	0.124	0.102	0.085	0.072	0.067
800	1.000	0.975	0.952	0.903	0.851	0.792	0.724	0.649	0.572	0.497	0.430	0.371	0.280	0.217	0.172	0.139	0.115	0.096	0.082	0.076
900	1.000	0.976	0.952	0.904	0.853	0.797	0.734	0.665	0.593	0.522	0.456	0.397	0.304	0.237	0.188	0.153	0.127	0.106	0.091	0.084
1000	1.000	0.976	0.952	0.904	0.855	0.801	0.742	0.677	0.609	0.542	0.478	0.420	0.325	0.255	0.204	0.167	0.138	0.116	0.099	0.092
1100	1.000	0.976	0.952	0.905	0.856	0.804	0.748	0.687	0.623	0.559	0.497	0.440	0.344	0.272	0.219	0.179	0.149	0.126	0.107	0.100
1200	1.000	0.976	0.952	0.905	0.857	0.807	0.753	0.695	0.634	0.573	0.513	0.457	0.362	0.288	0.233	0.192	0.160	0.135	0.116	0.107
1300	1.000	0.976	0.952	0.905	0.858	0.809	0.757	0.701	0.643	0.584	0.527	0.472	0.378	0.303	0.247	0.203	0.170	0.144	0.123	0.115
1400	1.000	0.976	0.952	0.906	0.859	0.811	0.760	0.707	0.651	0.595	0.539	0.486	0.392	0.317	0.259	0.214	0.180	0.153	0.131	0.122
1500	1.000	0.976	0.952	0.906	0.860	0.813	0.763	0.712	0.658	0.603	0.550	0.498	0.405	0.330	0.271	0.225	0.189	0.161	0.138	0.129
1600	1.000	0.976	0.952	0.906	0.861	0.814	0.766	0.716	0.664	0.611	0.559	0.508	0.417	0.342	0.282	0.235	0.198	0.169	0.145	0.135
1700	1.000	0.976	0.952	0.906	0.861	0.815	0.768	0.719	0.669	0.618	0.567	0.518	0.428	0.353	0.292	0.245	0.207	0.177	0.152	0.142
1800	1.000	0.976	0.952	0.906	0.862	0.816	0.770	0.722	0.673	0.624	0.574	0.526	0.438	0.363	0.302	0.254	0.215	0.184	0.159	0.148
1900	1.000	0.976	0.952	0.907	0.862	0.817	0.772	0.725	0.677	0.629	0.581	0.534	0.447	0.373	0.312	0.262	0.223	0.191	0.165	0.154
2000	1.000	0.976	0.952	0.907	0.863	0.818	0.773	0.728	0.681	0.634	0.587	0.541	0.455	0.382	0.320	0.271	0.230	0.198	0.172	0.160

Values of slenderness ratio $\lambda\,(=L_e/i)$

Equivalent L_e/b (for rectangular sections)

2.14.4 Slenderness ratio λ

The slenderness ratio of posts is given by the following general expression:

$$\lambda = \frac{\text{effective length}}{\text{least radius of gyration}} = \frac{L_e}{i}$$

For rectangular or square sections it may also be obtained from the following expression.

$$\lambda = \frac{\text{effective length}}{\text{least lateral dimension}} = \frac{L_e}{b}$$

The maximum slenderness ratio for members carrying dead and imposed loads is limited to either $L_e/i = 180$ or $L_e/b = 52$. Values greater than these limits indicate that a larger section is required.

Guidance on the effective length to be adopted, taking end restraint into consideration, is given in BS 5268 Table 21, reproduced here as Table 2.10.

2.14.5 Ratio of modulus of elasticity to compression stress

The ratio of modulus of elasticity to compression stress $E/\sigma_{c,\,par}$ must also be calculated to obtain K_{12} from Table 2.9. Here E is E_{min} for the timber grade, and $\sigma_{c,\,par}$ is $\sigma_{c,\,g,\,par} \times$ load duration factor K_3. Hence

$$\text{Ratio} = \frac{E_{min}}{\sigma_{c,\,g,\,par}K_3}$$

Table 2.10   Effective length of compression members (BS 5268 Part 2 1988 Table 21)

End conditions	Effective length / Actual length L_e/L
Restrained at both ends in position and in direction	0.7
Restrained at both ends in position and one end in direction	0.85
Restrained at both ends in position but not in direction	1.0
Restrained at one end in position and in direction and at the other end in direction but not in position	1.5
Restrained at one end in position and in direction and free at the other end	2.0

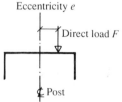

Figure 2.6   *Eccentrically loaded post*

2.14.6   Eccentrically loaded posts

In situations where the direct load is applied eccentrically, as shown in Figure 2.6, a bending moment will be induced equal to the applied load multiplied by the eccentricity:

$$\text{Eccentricity moment } M_e = Fe$$

The effect of this moment should be checked by ensuring first that the applied bending stress $\sigma_{m,a}$ is less than the permissible bending stress $\sigma_{m,adm}$, and then that the interaction quantity given by the following formula is less than unity:

$$\text{Interaction quantity} = \frac{\sigma_{m,a,par}}{\sigma_{m,adm,par}(1 - 1.5\sigma_{c,a,par}K_{12}/\sigma_e)} + \frac{\sigma_{c,a,par}}{\sigma_{c,adm,par}} \leqslant 1$$

where

$\sigma_{m,a,par}$ applied bending stress parallel to grain

$\sigma_{m,adm,par}$ permissible bending stress parallel to grain

σ_e Euler critical stress $= \pi^2 E_{min}/(L_e/i)^2$

$\sigma_{c,a,par}$ applied compression stress parallel to grain

$\sigma_{c,adm,par}$ permissible compression stress parallel to grain (including K_{12} factor)

2.14.7   Design summary for timber posts

The design procedure for axially loaded timber posts may be summarized as follows:

(a) Determine the effective length L_e dependent on end fixity.

(b) Calculate slenderness ratio λ from either of

$$\lambda = \frac{\text{effective length}}{\text{radius of gyration}} = \frac{L_e}{i} < 180$$

$$\lambda = \frac{\text{effective length}}{\text{least lateral dimension}} = \frac{L_e}{b} < 52$$

(c) Calculate the ratio of modulus of elasticity to compression stress $E_{min}/\sigma_{c,g,par}K_3$.

(d) Obtain K_{12} from Table 2.9 using λ and $E_{min}/\sigma_{c,g,par}K_3$ values.

(e) Calculate permissible compression stress:

$$\sigma_{c,adm,par} = \text{grade compression stress} \times K \text{ factors}$$
$$= \sigma_{c,g,par}K_3 K_{12} K_8$$

(where K_8 is used if applicable).

(f) Calculate applied stress and compare with permissible stress:

$$\sigma_{c,a,par} = \frac{\text{applied load}}{\text{area of section}} < \sigma_{c,adm,par}$$

Alternatively the permissible load can be calculated and compared with the applied load:

$$\text{Permissible load} = \sigma_{c,adm,par} \times \text{area} > \text{applied load}$$

If the post is eccentrically loaded, the following additional steps should be taken:

(g) Calculate eccentricity moment $M_e = \text{load} \times e$.

(h) Obtain grade bending stress $\sigma_{m,g,par}$ from table.

(i) Calculate permissible bending stress:

$$\sigma_{m,adm,par} = \sigma_{m,g,par}K_3 K_7$$

(j) Calculate applied bending stress and compare with permissible stress:

$$\sigma_{m,a,par} = \frac{M_e}{Z} < \sigma_{m,adm,par}$$

(k) Finally, check that the interaction quantity is less than unity:

$$\frac{\sigma_{m,a,par}}{\sigma_{m,adm,par}\{1 - [1.5\sigma_{c,a,par}K_{12}(L_e/i)^2/\pi^2 E_{min}]\}} + \frac{\sigma_{c,a,par}}{\sigma_{c,adm,par}} \leqslant 1$$

Note that throughout this procedure the wet exposure modification factors should be applied where necessary if the member is to be used externally.

Let us now look at some examples on the design of timber posts.

Example 2.5

What is the safe long term axial load that a 75 mm × 150 mm sawn GS grade hem–fir post can support if it is restrained at both ends in position and one end in direction, and its actual height is 2.1 m?

By reference to Table 2.10 the effective length L_e will be 0.85 times the actual length L. The effective length is used to calculate the slenderness ratio which, together with the ratio of modulus of elasticity to compression stress, is used to obtain the K_{12} factor from Table 2.9.

The slenderness ratio λ may be calculated from either of the following two expressions:

$$\lambda = \frac{\text{effective length}}{\text{radius of gyration}} = \frac{L_e}{i} = \frac{0.85 \times 2100}{21.7} = 82.25 < 180$$

$$\lambda = \frac{\text{effective length}}{\text{least lateral dimension}} = \frac{L_e}{b} = \frac{0.85 \times 2100}{75} = 23.8 < 52$$

Both values are therefore satisfactory. It should be noted that the radius of gyration value used in the first expression is the least radius of gyration about the $y-y$ axis, obtained from Table 2.4.

The ratio of modulus of elasticity to compression stress is calculated from the following expression:

$$\frac{E_{\min}}{\sigma_{c,g,par} K_3} = \frac{5800}{6.8 \times 1} = 852.94$$

The K_{12} factor is now obtained by interpolation from Table 2.9 as 0.495. This is used to adjust the grade compression stress and hence take account of the slenderness:

Grade compression stress $\sigma_{c,g,par} = 6.8 \, \text{N/mm}^2$

Permissible stress $\sigma_{c,adm,par} = \sigma_{c,g,par} K_3 K_{12} = 6.8 \times 1 \times 0.495 = 3.37 \, \text{N/mm}^2$

Permissible load = permissible stress × area of post
$$= 3.37 \times 11.3 \times 10^3 = 38 \times 10^3 \, \text{N} = 38 \, \text{kN}$$

Example 2.6

Design a timber post to support a medium term total axial load of 12.5 kN restrained in position but not in direction at both ends. The post is 2.75 m in height and GS grade redwood or whitewood is to be used.

The actual length $L = 2.75 \, \text{m} = 2750 \, \text{mm}$; the effective length $L_e = 1.0L$. Try 63 mm × 150 mm sawn section. The grade compression stress $\sigma_{c,g,par} = 6.8 \, \text{N/mm}^2$.

Calculate λ from either of the following:

$$\lambda = \frac{L_e}{i} = \frac{2750}{18.2} = 151 < 180$$

$$\lambda = \frac{L_e}{b} = \frac{2750}{63} = 43.65 < 52$$

Both values are satisfactory. Next,

$$\frac{E_{min}}{\sigma_{c,g,par}K_3} = \frac{5800}{6.8 \times 1.25} = 682.35$$

Thus from Table 2.9, $K_{12} = 0.168$. Finally, compare stresses:

Permissible compression stress:

$$\sigma_{c,adm,par} = \sigma_{c,g,par}K_3 K_{12} = 6.8 \times 1.25 \times 0.168 = 1.43 \, \text{N/mm}^2$$

Applied compression stress:

$$\sigma_{c,a,par} = \frac{\text{applied load}}{\text{section area}} = \frac{12.5 \times 10^3}{9.45 \times 10^3} = 1.32 \, \text{N/mm}^2 < 1.43 \, \text{N/mm}^2$$

Thus the section is adequate.

Alternatively the section may be checked by calculating the safe load it would sustain and comparing it with the applied load:

Safe load – permissible stress × section area

$$= 1.43 \times 9.45 \times 10^3 = 13.51 \times 10^3 \, \text{N} = 13.51 \, \text{kN} > 12.5 \, \text{kN}$$

Use 63 mm × 150 mm sawn GS grade redwood or whitewood post.

Example 2.7

An SS grade Scots pine post 2.5 m in height supports a total long term load of 40 kN applied 75 mm eccentric to its x–x axis as shown in Figure 2.7. Check the adequacy of a 100 mm × 250 mm sawn section if it is restrained at both ends in position and one end in direction.

Since the load is applied eccentrically, a bending moment will be developed for which the section must also be checked. The eccentricity moment about the x–x axis is given by

$$M_e = Fe = 40 \times 10^3 \times 75 = 3 \times 10^6 \, \text{N mm}$$

Calculate λ from either of the following:

$$\lambda = \frac{L_e}{i} = \frac{2500 \times 0.85}{28.9} = 73.53 < 180$$

$$\lambda = \frac{L_e}{b} = \frac{2500 \times 0.85}{100} = 21.25 < 52$$

Both values are satisfactory.

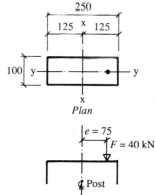

Figure 2.7 *Post loading and dimensions*

The grade compression stress $\sigma_{c,g,par} = 7.9 \text{ N/mm}^2$, and

$$\frac{E_{min}}{\sigma_{c,g,par}K_3} = \frac{6600}{7.9 \times 1} = 835.44$$

Thus from Table 2.9, $K_{12} = 0.553$. Next, compare stresses:

Permissible compression stress:

$$\sigma_{c,adm,par} = \sigma_{c,g,par}K_3K_{12} = 7.9 \times 1 \times 0.553 = 4.36 \text{ N/mm}^2$$

Applied compression stress:

$$\sigma_{c,a,par} = \frac{F}{A} = \frac{40 \times 10^3}{25 \times 10^3} = 1.6 \text{ N/mm}^2 < 4.36 \text{ N/mm}^2$$

Thus the section is satisfactory.

Having checked the effect of direct compression, the effect of the eccentricity moment must also be checked:

Grade bending stress $\sigma_{m,g,par} = 7.5 \text{ N/mm}^2$

Permissible bending stress:

$$\sigma_{m,adm,par} = \sigma_{m,g,par}K_3K_7 = 7.5 \times 1 \times 1.02 = 7.65 \text{ N/mm}^2$$

Applied bending stress:

$$\sigma_{m,a,par} = \frac{M_e}{Z_{xx}} = \frac{3 \times 10^6}{1040 \times 10^3} = 2.89 \text{ N/mm}^2 < 7.65 \text{ N/mm}^2$$

Again this is adequate.

Finally, the interaction quantity must be checked:

$$\frac{\sigma_{m,a,par}}{\sigma_{m,adm,par}\{1 - [1.5\sigma_{c,a,par}K_{12}(L_e/i)^2/\pi^2 E_{min}]\}} + \frac{\sigma_{c,a,par}}{\sigma_{c,adm,par}}$$

$$= \frac{2.89}{7.65\{1 - [1.5 \times 1.6 \times 0.553 \times (73.53)^2/\pi^2 \times 6600]\}} + \frac{1.6}{4.36}$$

$$= 0.425 + 0.367 = 0.792 < 1$$

Thus the 100 mm × 250 mm sawn section is adequate.

2.15   Load bearing stud walls

A cross-sectional plan through a typical stud wall is shown in Figure 2.8. For the purpose of design the studs may be regarded as a series of posts.

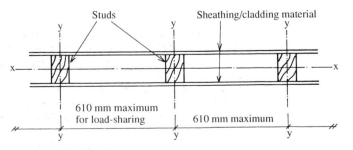

Figure 2.8   *Plan on a typical stud wall*

Provided that their centres do not exceed 610 mm they are considered to be load sharing, and hence the K_8 factor of 1.1 will apply.

The individual studs are usually taken to be laterally restrained about the y–y axis either by the sheathing/cladding material or by internal noggings or diagonal bracing. Hence their strength is calculated about the plane parallel to the wall, that is the x–x axis of the studs.

Since such walls are normally provided with a top and bottom rail, it is usual to consider that the loading is applied axially and that the ends are restrained in position but not in direction.

The following example will be used to illustrate stud wall design in relation to the previous procedure for posts.

Example 2.8

A stud wall panel is to be constructed within a maximum overall thickness of 100 mm, lined on both faces with 12.5 mm plasterboard. The panel is to be 2.4 m high overall including top and bottom rails with vertical studs placed at 600 mm centres, and is to be provided with intermediate horizontal noggings. Design suitable SC1 studs if the panel is to sustain a 7.5 kN per metre run long term load which includes the self-weight.

Panel load = 7.5 kN per metre run

Axial load per stud = $7.5 \times 0.6 = 4.5$ kN

The depth of timber is governed by the overall panel thickness less the plasterboard linings (see Figure 2.9). Thus the maximum stud depth $= 100 - (2 \times 12.5) = 75$ mm. In addition, the minimum practical breadth for fixing plasterboard is 38 mm.

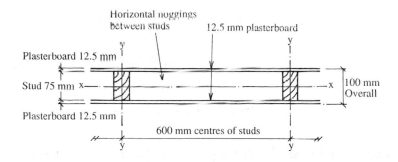

Figure 2.9   *Cross-sectional plan on stud wall*

It would be usual to specify regularized timber for wall studs. However, since only properties for sawn timber are given in this manual, a section from Table 2.4 will be selected for checking. Try the nearest practical size, that is 38 mm × 75 mm sawn timber studs.

As the studs are braced laterally by the horizontal noggings they can only buckle about their x–x axis. Therefore the i_x value or depth is used to calculate the slenderness ratio. Use either of the following expressions:

$$\lambda = \frac{L_e}{i_x} = \frac{2400 \times 1}{21.7} = 110.6 < 180$$

$$\lambda = \frac{L_e}{d} = \frac{2400 \times 1}{75} = 32 < 52$$

Both values are satisfactory.

The grade compression stress $\sigma_{c,g,par} = 3.5$ N/mm^2, and

$$\frac{E_{min}}{\sigma_{c,g,par} K_3} = \frac{4500}{3.5 \times 1} = 1285.71$$

Thus from Table 2.9, $K_{12} = 0.420$. Finally, compare stresses:

Permissible compression stress:

$$\sigma_{c,adm,par} = \sigma_{c,g,par} K_3 K_8 K_{12} = 3.5 \times 1 \times 1.1 \times 0.420 = 1.62 \text{ N/mm}^2$$

Applied compression stress:

$$\sigma_{c,a,par} = \frac{F}{A} = \frac{4.5 \times 10^3}{2.85 \times 10^3} = 1.58 \text{ N/mm}^2 < 1.62 \text{ N/mm}^2$$

Use 38 mm × 75 mm SC1 sawn timber studs.

2.16   Timber temporary works

Under this general heading may be included formwork for concrete, timber support work for excavations, or any other form of falsework giving temporary support to a permanent structure.

With respect to the actual structural elements comprising temporary works, these are basically either flexural or compression members whose design follows the procedures already explained. It should be appreciated, however, that the timber in temporary works will usually be in a wet exposure situation, and this must be taken into consideration in the design by applying the wet exposure modification factors where necessary.

The only other significant differences in approach relate to loading and matters of a practical nature. Reference should be made to the following British Standards for detailed guidance on particular aspects for the design of falsework and support to excavations:

BS 5975 1982   Code of practice for falsework.

BS 6031 1981   Code of practice for earthworks.

The implications of these codes will now be discussed in relation to two types of temporary works: formwork, and support work for excavations.

2.16.1   Formwork

Formwork is used to mould concrete into the desired shape and to provide temporary support to freshly poured concrete. Practical considerations play an important part in the conceptual design of formwork, particularly with respect to achieving economy by repetition and reuse. Perhaps the three most essential requirements for formwork are the following:

(a) It should be capable of carrying all working loads and pressures without appreciable deflection, during placing of the concrete.

(b) It should ideally be self-aligning, and all panels, bearers, props and other components should be capable of being easily assembled and dismantled in the desired sequence.

(c) The size of formwork components should be such that they are not too heavy to handle and will give repeated use without alteration.

Timber formwork may be designed directly in accordance with BS 5975, which contains information on all the necessary stresses and modification factors. The general principles employed are similar to those in BS 5268, with slight variations because of the temporary nature of falsework. Reference still needs to be made to BS 5268 for information on the various properties of the timber sections.

BS 5975 recommends that SC3 timber should be the minimum quality adopted for falsework, and it gives wet exposure grade stresses for SC3, SC4 and SC5 timber.

One essential difference between BS 5975 and BS 5268 is the load duration factor K_3 for timber used in falsework. Values of K_3 for falsework are given in Table 5 of BS 5975, reproduced here as Table 2.11. The periods given in the table relate to construction times; it should be noted that these are cumulative over the life of the timber, unless there is a time lapse between load periods at least equal to the time the timber was previously loaded.

Table 2.11   Modification factor K_3 for duration of load on falsework
(BS 5975 1982 Table 5)

Duration of loading	K_3
1 year	1.2
1 month	1.3
1 week	1.4

Compared with BS 5268, the permissible shear stresses for timber falsework given in BS 5975 are increased by a factor of 1.5 because of the temporary nature of the loading.

Deflection of formwork is an important factor since it directly affects the appearance of the finished concrete face. It is therefore recommended that formwork deflection be limited to the lesser of $0.003 \times$ span or 3 mm.

Reference should be made to BS 5975 for guidance on the loads that formwork must sustain. These should include an allowance for construction operations together with all permanent loading from the concrete and the self-weight.

As an alternative to designing formwork for a specific purpose, standard solutions are available from proprietary manufacturers. A standard solution usually involves the selection of suitable components for which the design information is presented in a tabular form by the manufacturer. Commonly available are systems for waffle slab construction, flat slab soffits, wall faces, beams and columns. The systems may utilize steel and aluminium as well as timber.

The following example illustrates a typical design for formwork in accordance with BS 5975.

Example 2.9

The arrangement for a system of formwork to support a 125 mm thick reinforced concrete slab is shown in Figure 2.10. The following data are given:

Load due to weight of reinforced concrete: 24 kN/m^3
Load due to self-weight of timber sheeting: 0.1 kN/m^2
Load due to self-weight of joists: 0.12 kN/m^2
Imposed load due to construction work: 1.5 kN/m^2

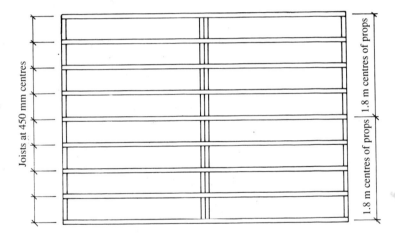

Plan

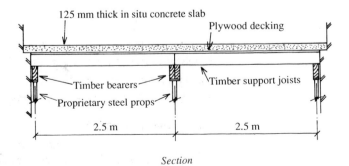

Section

Figure 2.10 *Formwork supporting a reinforced concrete slab*

The timber will be SC3, having the following wet exposure grade stresses obtained by multiplying the dry stresses by the relevant K_2 factor from Table 2.3:

Bending stress parallel to grain $\sigma_{m,g,par} = 4.24\ \text{N/mm}^2$

Shear stress parallel to grain r_g $= 0.603\ \text{N/mm}^2$

Mean modulus of elasticity E_{mean} $= 7040\ \text{N/mm}^2$

The following modification factors will apply:

Wet exposure geometrical factor K_1 from Table 2.5:
  For cross-sectional area: 1.04
  For section modulus: 1.06
  For second moment of area: 1.08

Load duration factor K_3 for 1 week (from Table 2.11) = 1.4

Load sharing factor $K_8 = 1.1$

Depth factor K_7 as applicable to size

Maximum depth to breadth ratio = 5

Design a typical support joist spanning 2.5 m at 450 mm centres.

Loading

Dead load: concrete $24 \times 0.125 = 3.0$
        timber sheeting $= 0.1$
        timber joists $= \underline{0.12}$
            $3.22\ \text{kN/m}^2$

Imposed load: $1.5\ \text{kN/m}^2$

Combined load: dead   3.22
        imposed $\underline{1.50}$
            $4.72\ \text{kN/m}^2$

UDL per joist $= 4.72 \times 2.5 \times 0.45 = 5.31\ \text{kN}$

Bending

$$M = \frac{WL}{8} = \frac{5.31 \times 2.5}{8} = 1.66\ \text{kN m} = 1.66 \times 10^6\ \text{N mm}$$

Wet exposure grade bending stress $\sigma_{m,g,par} = 4.24\ \text{N/mm}^2$

K_1 wet exposure section modulus factor $= 1.06$

K_3 load duration factor (1 week) from Table 2.11 $= 1.4$

K_8 load sharing factor $= 1.1$

K_7 depth factor is unknown at this stage

Approximate Z_{xx} required

$$= \frac{M}{\sigma_{m,g,par} K_1 K_3 K_8} = \frac{1.66 \times 10^6}{4.24 \times 1.06 \times 1.4 \times 1.1} = 239\,837\ \text{mm}^3 = 240 \times 10^3\ \text{mm}^3$$

Maximum depth to breadth ratio for lateral stability is 5

Calculate the approximate I_{xx} required to satisfy bending deflection alone. The permissible deflection is the lesser of $0.003 \times \text{span} = 0.003 \times 2500 = 7.5\,\text{mm}$ or $3\,\text{mm}$, so will be $3\,\text{mm}$. From $\delta_m = (5/384)(WL^3/EI)$ and $\delta_p = 3\,\text{mm}$,

Approximate I_{xx} required

$$= \frac{5}{384} \times \frac{5.31 \times 10^3 \times 2500^3}{7040 \times 3} = 51\,151\,622\,\text{mm}^4 = 51.2 \times 10^6\,\text{mm}^4$$

This can be divided by the wet exposure K_1 factor to give the second moment of area:

$$\text{Final } I_{xx} \text{ required} = \frac{51.2 \times 10^6}{1.08} = 47.4 \times 10^6\,\text{mm}$$

From Table 2.4, for a $63\,\text{mm} \times 225\,\text{mm}$ sawn joist $Z_{xx} = 532 \times 10^3\,\text{mm}^3$ and $I_{xx} = 59.8 \times 10^6\,\text{mm}^4$.

Deflection

Actual deflection $\delta_a = \delta_m + \delta_v = \dfrac{5}{384}\dfrac{WL^3}{EI} + \dfrac{19.2M}{AE}$

The relevant wet exposure K_1 factors should be applied to the area and I values in this expression to give

$$\delta_a = \frac{5}{384}\frac{WL^3}{EIK_1} + \frac{19.2M}{K_1 AE}$$

$$= \frac{5}{384} \times \frac{5.31 \times 10^3 \times 2500^3}{7040 \times 59.8 \times 10^6 \times 1.08} + \frac{19.2 \times 1.66 \times 10^6}{1.04 \times 14.2 \times 10^3 \times 7040}$$

$$= 2.38 + 0.31 = 2.69\,\text{mm} < 3\,\text{mm}$$

This section is adequate.

Shear unnotched

Maximum shear $F_v = \dfrac{\text{UDL}}{2} = \dfrac{5.31}{2} = 2.66\,\text{kN} = 2.66 \times 10^3\,\text{N}$

Wet exposure $r_g = 0.603\,\text{N/mm}^2$

$r_{adm} = r_g K_3 K_8 = 0.603 \times 1.4 \times 1.1 = 0.929\,\text{N/mm}^2$

For falsework the r_{adm} may be increased by a further factor of 1.5 in accordance with BS 5975:

$r_{adm} = 0.929 \times 1.5 = 1.39\,\text{N/mm}^2$

$$r_a = \frac{3}{2}\frac{F_v}{K_1 A} = \frac{3}{2} \times \frac{2.66 \times 10^3}{1.04 \times 14.2 \times 10^3} = 0.27\,\text{N/mm}^3 < 1.39\,\text{N/mm}^2$$

Conclusion

Use $63\,\text{mm} \times 225\,\text{mm}$ SC3 sawn joists.

The size of the timber bearers would be determined in a similar manner and suitable plywood decking would be chosen to resist bending and deflection. Proprietary steel props would be used to support the bearers, selected by reference to a manufacturer's safe load tables.

2.16.2   Support work for excavations

The general term 'timbering' is often used to describe any form of temporary support work for excavations. It may however be composed of either timber or steel, or a combination of the two. The choice from the different methods and combinations available will be dependent on a number of factors, such as soil conditions and the plant to be used.

Methods of excavating trenches, pits and shafts in various types of ground and methods of providing temporary support to the sides are described in BS 6031.

The fundamental requirements of any support work may be summarized as follows:

(a) It should provide safe working conditions.
(b) It should allow both the excavation and the construction of the permanent work to be carried out efficiently.
(c) It should be capable of being easily and safely removed after completion of the permanent work.

Typical examples of timbering for shallow trench excavations are illustrated in Figures 2.11 and 2.12. For deep excavations where standard timber sizes are impractical, steel trench sheeting or driven interlocking steel sheet piles are normally used.

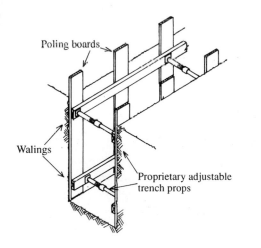

Figure 2.11   *Trench support for excavation in firm ground*

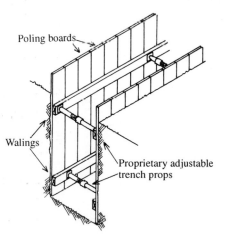

Figure 2.12   *Trench support for excavation in loose ground*

Practical considerations rather than pure structural design requirements often dictate the size of the timber sections that are adopted. Health and safety regulations and other codes require that an adequate supply of timber or other support material be kept available for use in excavations. Therefore the sizes that the designer has to work with may already be decided. In such cases he will advise on the spacing of the supports based on the structural capacity of the sections.

Since the actual design of the timber elements used to support excavations follows the BS 5268 procedures already described, only an appreciation of how the loads acting on trench supports are derived will be given here.

Consider the section through the typical trench support system shown in Figure 2.13. Forces exerted by the retained earth are transmitted from one side of the excavation to the other by walings and horizontal props. The vertical poling boards and horizontal walings are subject to bending for the distance they span between the trench props. Provided that the supporting members are adequately designed, a state of equilibrium will be maintained between the two sides.

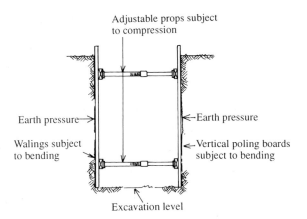

Figure 2.13 *Section through typical trench support system*

The loads acting on the timbering result from earth pressure, and for non-cohesive soils they may be calculated using Rankine's theory. In addition a pressure due to surcharge from any imposed loading adjacent to the trench may have to be allowed for. The pressure diagram for such loading is shown in Figure 2.14, and the relevant formulae for calculating the resultant forces acting on the timbering are as follows:

Maximum retained earth pressure at excavation level:

$$q = Wh \left[\frac{1 - \sin \theta}{1 + \sin \theta} \right]$$

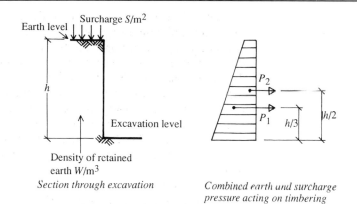

Section through excavation

Combined earth and surcharge
pressure acting on timbering

Figure 2.14   *Pressure acting on trench supports*

Maximum resultant thrust:

$$P_1 = \frac{qh}{2} = \frac{Wh^2}{2}\left[\frac{1-\sin\theta}{1+\sin\theta}\right]$$

Resultant thrust from any surcharge:

$$P_2 = Sh\left[\frac{1-\sin\theta}{1+\sin\theta}\right]$$

where

W density of retained material
h depth of excavation
θ angle of repose or angle of internal friction of retained material
S surcharge loading per unit area

Having obtained the forces acting on the timbering, the forces and bending moments induced in the members can be calculated and their size determined. In this context it is important to remember that the boarding and waling sections will be subject to bending about their minor axis, that is the $y–y$ axis.

Owing to the repetitive nature of trench support work, various combinations of the basic components have been developed into patented systems. These are available from a number of suppliers who are usually also able to provide a design service.

2.17   References

BS 4978 1988 Specification for softwood grades for structural use.
BS 5268 Structural use of timber.
    Part 2 1988 Code of practice for permissible stress design, materials and workmanship.
BS 5975 1982 Code of practice for falsework.
BS 6031 1981 Code of practice for earthworks.

Timber Designers' Manual, 2nd edn.
J. A. Baird and E. C. Ozelton. BSP Professional Books, 1984.

TRADA Wood information sheets:
Section 1 Sheet 25 Introduction to BS 5268: Part 2.
Section 1 Sheet 26 Supplying timber to BS 5268: Part 2.
Section 1 Sheet 28 Timber and wood based sheet materials for floors.
Section 2/3 Sheet 15 Basic sizes of softwoods and hardwoods.

For further information contact:

Timber Research and Development Association (TRADA), Stocking Lane, Hughenden Valley, High Wycombe, Buckinghamshire, HP14 4ND.

3 Concrete elements

3.1 Structural design of concrete

Guidance on the use of concrete in building and civil engineering structures, other than for bridges and liquid retaining structures, is given in BS 8110 'Structural use of concrete'. This is divided into the following three parts:

Part 1 Code of practice for design and construction.
Part 2 Code of practice for special circumstances.
Part 3 Design charts for singly reinforced beams, doubly reinforced beams and rectangular columns.

The design of reinforced concrete dealt with in this manual is based on Part 1, which gives recommendations for concrete elements contained in routine construction. Design charts presented in Part 3 will also be examined for comparison with the results obtained from the guidance given in Part 1.

Whilst Part 1 covers the majority of structural applications encountered in everyday design, circumstances may arise that require further assessment, such as torsional or other less common analyses. Part 2, which is complementary to Part 1, gives recommendations for such special circumstances.

For information on all aspects of bridge design, reference should be made to BS 5400 'Steel, concrete and composite bridges'. Similarly for the design of liquid retaining structures other than dams, reference is made to BS 8007 'Design of concrete structures for retaining aqueous liquids'. These are beyond the scope of this manual.

3.2 Symbols

BS 8110 adopts a policy of listing symbols at the beginning of each subsection. In this context care needs to be exercised since certain symbols appear in more than one place with subtle differences in definition.

Those symbols that are relevant to this manual are listed below and for ease of reference some are repeated under more than one heading:

General

f_{cu} characteristic strength of concrete
f_y characteristic strength of reinforcement
G_k characteristic dead load
Q_k characteristic imposed load
W_k characteristic wind load
SLS serviceability limit state
ULS ultimate limit state

γ_f partial safety factor for load
γ_m partial safety factor for strength of materials

Section properties

A_c net cross-sectional area of concrete in a column
A_s area of tension reinforcement
A_{sc} area of main vertical reinforcement
b width of section
b_c breadth of compression face of a beam
b_v breadth of section used to calculate the shear stress
d effective depth of tension reinforcement
h overall depth of section
x depth to neutral axis
z lever arm

Bending

A_s area of tension reinforcement
b width of section
d effective depth of tension reinforcement
f_{cu} characteristic strength of concrete
f_y characteristic strength of reinforcement
K coefficient obtained from design formula for rectangular beams
K' 0.156 when redistribution of moments does not exceed 10 per cent
M design ultimate resistance moment; or
M_u design ultimate bending moment due to ultimate loads
x depth to neutral axis
z lever arm

Deflection

b width of section
d effective depth of tension reinforcement
f_y characteristic strength of reinforcement
M design ultimate bending moment at centre of the span or, for a cantilever, at the support

Shear

A_s area of tension reinforcement
A_{sb} cross-sectional area of bent-up bars
A_{sv} total cross-section of links at the neutral axis
b_v breadth of section used to calculate the shear stress
d effective depth of tension reinforcement
f_{cu} characteristic strength of concrete
f_{yv} characteristic strength of links (not to exceed 460 N/mm^2)
s_b spacing of bent-up bars
s_v spacing of links along the member
V design shear force due to ultimate loads

V_b design shear resistance of bent-up bars
v design shear stress at a cross-section
v_c design concrete shear stress (from BS 8110 Table 3.9)
θ angle of shear failure plane from the horizontal
α angle between a bent-up bar and the axis of a beam
β angle between the compression strut of a system of bent-up bars and the axis of the beam

Compression

A_c net cross-sectional area of concrete in a column
A_{sc} area of vertical reinforcement
b width of column
f_{cu} characteristic strength of concrete
f_y characteristic strength of reinforcement
h depth of section
l_e effective height
l_{cx} effective height in respect of major axis
l_{cy} effective height in respect of minor axis
l_o clear height between end restraints
N design ultimate axial load on a column

3.3 Design philosophy

The design of timber in Chapter 2 was based on permissible stress analysis, whereas the design analysis for concrete employed in BS 8110 is based on limit state philosophy. Its object is to achieve an acceptable probability that the structure being designed will not become unfit for its intended purpose during its expected life. Therefore the various ways in which a structure could become unfit for use are examined.

The condition of a structure when it becomes unfit for use or unserviceable is called a limit state. This can by definition be further subdivided into the following two categories:

(a) Ultimate limit state (ULS)

(b) Serviceability limit state (SLS).

3.3.1 Ultimate limit state

If a ULS is reached, collapse of the member or structure will occur. Therefore the design must examine all the ULSs likely to affect a particular member. Some of the ULSs that may have to be considered are as follows:

(a) ULS due to bending

(b) ULS due to shear

(c) ULS due to direct compression or tension

(d) ULS due to overturning.

3.3.2 Serviceability limit state

If an SLS is reached the appearance of the member or structure will be disrupted. Whilst this will not cause collapse it may render the member unfit for its intended service use. Some of the SLSs that may have to be considered are as follows:

(a) SLS due to deflection: this should not adversely affect the appearance of the structure.
(b) SLS due to cracking: this should not adversely affect the appearance or the durability of the structure. For example, excessive cracks would allow the ingress of moisture with subsequent corrosion and/or frost damage.
(c) SLS due to vibration: this should not produce structural damage or cause discomfort or alarm to occupants of the building. Special precautions may be necessary to isolate the source of such vibration.

Other serviceability considerations that may have to be taken into account in the design of a particular member or structure are durability, fatigue, fire resistance and lightning.

Having identified the various limit states, the basic design procedure to ensure that they are not exceeded may be summarized as follows.

3.3.3 Limit state basic design procedure

When designing a particular concrete element it is usual to first ensure that the ULS is not exceeded and then to check that the relevant SLSs are also satisfied.

In order to ensure that the ULS is not exceeded, safety factors are applied as discussed in the next section.

The serviceability requirements for routine design are usually met by compliance with certain dimensional ratios or detailing rules given in BS 8110 Part 1. They will be referred to later in the relevant sections of this chapter. If it were considered necessary to examine the deflection or cracking SLSs in more detail then reference may be made to the more rigorous method of analysis given in BS 8110 Part 2.

3.4 Safety factors

In previous codes of practice the design of reinforced concrete members was based on either elastic theory or load factor theory. The fundamental difference between the two methods is the application of safety factors: for elastic analysis they were applied to the material stresses, and for load factor analysis they were applied indirectly to the loads.

Limit state philosophy acknowledges that there can be variation in both the loads and the materials. Therefore in limit state analysis, partial safety factors are applied separately to both the loads and the material stresses.

3.5 Loads

It is accepted in limit state philosophy that the loads in practice may vary from those initially assumed. Therefore the basic load is adjusted by a partial safety factor to give the ultimate design load. Each of these will be discussed in turn.

3.5.1. Characteristic loads

These are the basic loads that may be applicable to a particular member or structure and are defined as follows:

Characteristic dead load G_k   The weight of the structure complete with finishes, fixtures and partitions, obtained from BS 648 'Schedule of weights of building materials'.

Characteristic imposed load Q_k   The live load produced by the occupants and usage of the building, obtained from BS 6399 'Design loading for buildings', Part 1 for floors or Part 3 for roofs.

Characteristic wind load W_k   The wind load acting on the structure, obtained from CP 3 Chapter V Part 2 'Wind loads', which will eventually become Part 2 of BS 6399.

3.5.2 Partial safety factors for load

In practice the applied load may be greater than the characteristic load for any of the following reasons:

(a) Calculation errors

(b) Constructional inaccuracies

(c) Unforeseen increases in load.

To allow for these the respective characteristic loads are multiplied by a partial safety factor γ_f to give the ultimate design load appropriate to the limit state being considered. That is,

$$\text{Ultimate design load} = \gamma_f \times \text{characteristic load}$$

Values of γ_f for various load combinations are given in BS 8110 Table 2.1, reproduced here as Table 3.1.

Table 3.1   Load combinations and values of γ_f for the ultimate limit state (BS 8110 Part 1 1985 Table 2.1)

Load combination	Dead load		Imposed load		Earth and water pressure	Wind load
	Adverse	Beneficial	Adverse	Beneficial		
Dead and imposed (and earth and water pressure)	1.4	1.0	1.6	0	1.4	—
Dead and wind (and earth and water pressure)	1.4	1.0	—	—	1.4	1.4
Dead and wind and imposed (and earth and water pressure)	1.2	1.2	1.2	1.2	1.2	1.2

3.5.3   Ultimate design load

The ultimate design load acting on a member will be the summation of the relevant characteristic load combinations multiplied by their respective partial safety factors. Thus the ultimate design load for the combination of dead and imposed loads would be expressed as follows:

$$\text{Ultimate design load } F \text{ dead} + \text{imposed} = \gamma_f G_k + \gamma_f Q_k$$
$$= 1.4G_k + 1.6Q_k$$

The following examples illustrate the computation of loads for limit state design. They may be compared with the examples in Chapter 1 for permissible stress design.

Example 3.1

A series of 400 mm deep × 250 mm wide reinforced concrete beams spaced at 5 m centres and spanning 7.5 m support a 175 mm thick reinforced concrete slab as shown in Figure 3.1. If the imposed floor loading is 3 kN/m² and the load induced by the weight of concrete is 24 kN/m³, calculate the total ULS loading condition for the slab and the beams.

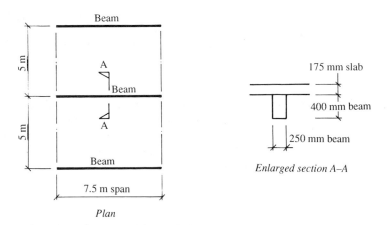

Figure 3.1   *Arrangement of beams*

Slab

The ULS loading condition for the slab will be the dead plus imposed combination:

Dead load G_k from 175 mm slab $= 24 \times 0.175 = 4.2\,\text{kN/m}^2$
Imposed load $Q_k = 3\,\text{kN/m}^2$
Total ULS loading $= \gamma_f G_k + \gamma_f Q_k = 1.4 \times 4.2 + 1.6 \times 3 = 10.68\,\text{kN/m}^2$

Beam

The ULS loading condition for the beams will be a UDL consisting of the slab dead plus imposed combination together with the load due to the self-weight of the beam:

Slab ULS total UDL $= 10.68 \times 7.5 \times 5 = 400.5\,\text{kN}$

Beam G_k self-weight UDL $= 24 \times 0.4 \times 0.25 \times 7.5 = 18\,\text{kN}$

ULS beam UDL $= \gamma_f G_k = 1.4 \times 18 = 25.2\,\text{kN}$

Total ULS beam UDL $= 400.5 + 25.2 = 425.7\,\text{kN}$

These total loads would be used to design the slab and beams for the ULSs of bending and shear.

Example 3.2

A 3 m high reinforced concrete column supports a 700 kN characteristic dead load and a 300 kN characteristic imposed load. Calculate the total ULS design load for the column if it is 300 mm $\times$ 250 mm in cross-section and the load due to the weight of concrete is $24\,\text{kN/m}^3$.

Applied dead load $G_k = 700\,\text{kN}$

Self-weight dead load $G_k = 24 \times 3 \times 0.3 \times 0.25 = 5.4\,\text{kN}$

Applied imposed load $Q_k = 300\,\text{kN}$

Total ULS design load $= \gamma_f G_k + \gamma_f Q_k$
$$= 1.4 \times 700 + 1.4 \times 5.4 + 1.6 \times 300 = 1467.56\,\text{kN}$$

This load would be used to design the column for the ULS of axial compression.

3.6 Material properties

The strength of the materials actually used in construction can vary from the specified strength for a number of reasons. Therefore in ULS design the basic or characteristic strength of a material is modified by a partial safety factor to give the ultimate design strength. This is explained in more detail below.

3.6.1 Characteristic strength of materials

BS 8110 adopts the criterion that no more than 5 per cent of a sample batch should have less than a specified strength. This strength is called the characteristic strength, denoted by f_{cu} for the concrete and f_y for the steel reinforcement.

The test results used for specifying the characteristic strengths of reinforced concrete materials are the cube strengths of concrete and the yield or proof strength of steel reinforcement.

Table 9 of BS 5328 'Concrete' Part 1, 'Guide to specifying concrete', reproduced here as Table 3.2, lists the characteristic strengths for various grades of concrete. These are in fact the cube strengths of the concrete at 28 days. The yield strength of reinforcement is given in BS 8110 Table 3.1, reproduced here as Table 3.3.

Table 3.2   Concrete compressive strength (BS 5328 Part 1 1990 Table 9)

Concrete grade	Characteristic compressive strength at 28 days $(N/mm^2 = MPa)$
C7.5	7.5
C10	10.0
C12.5	12.5
C15	15.0
C20	20.0
C25	25.0
C30	30.0
C35	35.0
C40	40.0
C45	45.0
C50	50.0
C55	55.0
C60	60.0

Table 3.3   Strength of reinforcement (BS 8110 Part 1 1985 Table 3.1)

Designation	Specified characteristic strength f_y (N/mm^2)
Hot rolled mild steel	250
High yield steel (hot rolled or cold worked)	460

3.6.2   Partial safety factors for materials

For the analysis of reinforced concrete elements the design strength of the concrete and the steel reinforcement is obtained by dividing their characteristic strength by a partial safety factor γ_m. This factor is to take account of differences that may occur between laboratory and on-site values. Such differences could be caused by any of the following:

Concrete

Segregation during transit
Dirty casting conditions
Inadequate protection during curing
Inadequate compaction of concrete.

Reinforcement

Wrongly positioned reinforcement
Distorted reinforcement
Corroded reinforcement.

Values of γ_m for the ULS are given in BS 8110 Table 2.2, which is reproduced here as Table 3.4.

Table 3.4   Values of γ_m for the ultimate limit state (BS 8110 Part 1 1985 Table 2.2)

Reinforcement	1.15
Concrete in flexure or axial load	1.50
Shear strength without shear reinforcement	1.25
Bond strength	1.4
Others (e.g. bearing stress)	$\geqslant 1.5$

3.6.3   Ultimate design strength of materials

The ultimate design strength of a material is obtained by dividing its characteristic strength by the appropriate partial safety factor referred to in Section 3.6.2:

$$\text{Ultimate design strength of concrete} = \frac{f_{cu}}{1.5} = 0.67 f_{cu}$$

$$\text{Ultimate design strength of reinforcement} = \frac{f_y}{1.15} = 0.87 f_y$$

It is important to appreciate that the formulae and design charts given in BS 8110 have been derived with the relevant partial safety factors for strength included. Therefore it is only necessary for the designer to insert the relevant characteristic strength values f_{cu} or f_y in order to use the formulae and charts.

3.7   Practical considerations for durability

Before proceeding to the actual structural design of concrete elements, a number of important practical considerations related to durability are worthy of mention since they can influence the size of members.

Durable concrete should perform satisfactorily in its intended environment for the life of the structure. To achieve durable concrete it is necessary to consider several interrelated factors at different stages in both the design and construction phases. Guidance is given in BS 8110 on various factors that influence reinforced concrete durability. They include:

(a) Shape and bulk of concrete

(b) Amount of concrete cover to reinforcement

(c) Environmental conditions to which the concrete will be exposed

(d) Cement type

(e) Aggregate type

(f) Cement content and water to cement ratio

(g) Workmanship necessary to attain full compaction and effective curing of the concrete.

Factors (a) and (b) must be considered at the design stage because they influence the member size and the location of the reinforcement. These are therefore discussed in more detail below. The remaining factors listed may be catered for by including suitable clauses in the specification and by adequate site management.

3.7.1 Shape and bulk of concrete

If the concrete will be exposed when the building is finished, adequate thought should be given at the design stage to its shape and bulk to prevent the ingress of moisture. The shape should be detailed to encourage natural drainage and hence avoid standing water.

3.7.2 Concrete cover to reinforcement

All reinforcement must be provided with sufficient cover to avoid corrosion and guard against distortion in the event of fire. The amount of cover to protect against fire is discussed in Section 3.7.3.

The amount of cover necessary to protect reinforcement against corrosion depends on both the exposure conditions that prevail and the quality of concrete used. BS 8110 Table 3.2 defines exposure conditions, and Table 3.4 gives the nominal cover to be provided with respect to the concrete quality. These tables are reproduced here as Tables 3.5 and 3.6 respectively.

Table 3.5   Exposure conditions (BS 8110 Part 1 1985 Table 3.2)

Environment	Exposure conditions
Mild	Concrete surfaces protected against weather or aggressive conditions
Moderate	Concrete surfaces sheltered from severe rain or freezing whilst wet Concrete subject to condensation Concrete surfaces continuously under water Concrete in contact with non-aggressive soil (see class 1 of Table 6.1 of BS 8110)*
Severe	Concrete surfaces exposed to severe rain, alternate wetting and drying, or occasional freezing or severe condensation
Very severe	Concrete surfaces exposed to sea water spray, de-icing salts (directly or indirectly), corrosive fumes or severe freezing conditions whilst wet
Extreme	Concrete surfaces exposed to abrasive action, e.g. sea water carrying solids or flowing water with pH $\leqslant 4.5$ or machinery or vehicles

* For aggressive soil conditions see clause 6.2.3.3 of BS 8110.

Table 3.6   Nominal cover to all reinforcement (including links) to meet durability requirements (BS 8110 Part 1 1985 Table 3.4)

Conditions of exposure‡	Nominal cover (mm)				
Mild	25	20	20*	20*	20*
Moderate	—	35	30	25	20
Severe	—	—	40	30	25
Very severe	—	—	50†	40†	30
Extreme	—	—	—	60†	50
Maximum free water/cement ratio	0.65	0.60	0.55	0.50	0.45
Minimum cement content (kg/m³)	275	300	325	350	400
Lowest grade of concrete	C30	C35	C40	C45	C50

* These covers may be reduced to 15 mm provided that the nominal maximum size of aggregate does not exceed 15 mm.

† Where concrete is subject to freezing whilst wet, air-entrainment should be used (see clause 3.3.4.2 of BS 8110).

‡ For conditions of exposure see Table 3.5 of this chapter.

Note 1: This table relates to normal-weight aggregate of 20 mm nominal maximum size.

Note 2: For concrete used in foundations to low rise construction (see clause 6.2.4.1 of BS 8110).

Two points should be noted. First, the cover stipulated is that to all reinforcement including any links. Secondly, the values are nominal and therefore under certain circumstances may have to be increased.

The amount of cover should also comply with recommendations given in BS 8110 relating to bar size, to aggregate size and to situations where the concrete is cast against uneven surfaces. It must also allow for any surface treatment, such as bush hammering, that would reduce the nominal thickness.

A summary of the requirements for cover is given in Table 3.7, and typical examples are illustrated in Figure 3.2.

Table 3.7   Summary of cover requirements (other than for fire resistance): cover to any bar, including links, is the greatest of the relevant values

Circumstances	Cover
Generally	Nominal value from Table 3.6
Relative to aggregate	Size of coarse aggregate
Resulting cover to single main bars	Bar diameter
Resulting cover to bundles of main bars	Bar diameter equivalent to area of group
Concrete cast against earth	75 mm
Concrete cast against blinding	40 mm

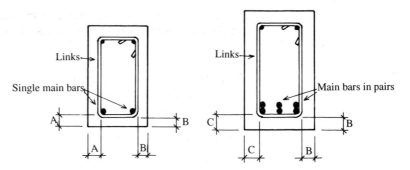

A = Cover to a single main bar ≮ bar diameter
B = Nominal cover to links ≮ value from Table 3.6 ≮ aggregate size
C = Cover to group of main bars ≮ bar diameter equivalent to area of group

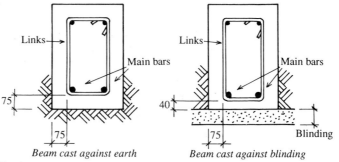

Beam cast against earth   *Beam cast against blinding*

Note:
For simplicity only beams have been used to illustrate the requirements for cover although similar requirements apply to other concrete members.

Figure 3.2 *Typical examples of cover to reinforcement*

3.7.3 Fire resistance

The fire resistance of a reinforced concrete member is dependent upon the cover to reinforcement, the type of aggregate that is used and the minimum dimensions of the member.

Nominal cover provided for protection against corrosion may, in certain circumstances, not suffice as fire protection. Reference should be made to BS 8110 Part 1 Table 3.5 and Figure 3.2 for the amount of cover and minimum member dimensions to satisfy fire resistance requirements.

Further guidance on design for fire, including information on surface treatments, is given in Section 4 of BS 8110 Part 2.

3.8 Flexural members

Flexural members are those subjected to bending, for example beams and slabs. Primarily the same procedure appertains to the design of both, although there are certain subtle differences. The design of beams will therefore be studied first and then compared with the design of slabs.

3.9 Beams

There are a number of dimensional requirements and limitations applicable to concrete beams which the designer needs to consider since they can affect the design:

(a) Effective span of beams
(b) Deep beams
(c) Slender beams
(d) Main reinforcement areas
(e) Minimum spacing of reinforcement
(f) Maximum spacing of reinforcement.

Certain other aspects such as bond, anchorage, and if applicable the curtailment and lap lengths of reinforcement, require consideration at the detailing stage.

The main structural design requirements for which concrete beams should be examined are as follows:

(a) Bending ULS
(b) Cracking SLS
(c) Deflection SLS
(d) Shear ULS.

Let us now consider how each of these dimensional and structural requirements influences the design of beams.

3.9.1 Effective span of beams

The effective span or length of a simply supported beam may be taken as the lesser of:

(a) The distance between the centres of bearing
(b) The clear distance between supports plus the effective depth d.

The effective length of a cantilever should be taken as its length to the face of the support plus half its effective depth d.

3.9.2 Deep beams

Deep beams having a clear span of less than twice their effective depth d are outside the scope of BS 8110. Reference should therefore be made to specialist literature for the design of such beams.

3.9.3   Slender beams

Slender beams, where the breadth of the compression face b_c is small compared with the depth, have a tendency to fail by lateral buckling. To prevent such failure the clear distance between lateral restraints should be limited as follows:

(a) For simply supported beams, to the lesser of $60b_c$ or $250b_c^2/d$

(b) For cantilevers restrained only at the support, to the lesser of $25b_c$ or $100b_c^2/d.$

These slenderness limits may be used at the start of a design to choose preliminary dimensions. Thus by relating the effective length of a simply supported beam to $60b_c$, an initial breadth can be derived. This can then be substituted in the bending design formula, given in Section 3.9.7, and an effective depth d determined. Finally this can be compared with the second slenderness limit of $250b_c^2/d$.

Example 3.3

A simply supported beam spanning 8 m is provided with effective lateral restraints at both ends. If it has an effective depth of 450 mm, what breadth would be satisfactory?

To avoid lateral buckling failure the distance between lateral restraints should be the lesser of $60b_c$ or $250b_c^2/d$. Hence either

$$\text{Effective span} = 60b_c$$
$$8000 = 60b_c$$
$$b_c = \frac{8000}{60} = 133 \text{ mm}$$

or

$$\text{Effective span} = \frac{250b_c^2}{d}$$
$$8000 = \frac{250b_c^2}{450}$$
$$b_c = \sqrt{\left(\frac{8000 \times 450}{250}\right)} = 120 \text{ mm}$$

Hence the minimum breadth of beam to avoid lateral buckling would have to be 133 mm.

3.9.4   Main reinforcement areas

Sufficient reinforcement must be provided in order to control cracking of the concrete. Therefore the minimum area of tension reinforcement in a beam should not be less than the following amounts:

(a) 0.24 per cent of the total concrete area, when $f_y = 250 \text{ N/mm}^2$

(b) 0.13 per cent of the total concrete area, when $f_y = 460 \text{ N/mm}^2$.

To ensure proper placing and compaction of concrete around reinforcement, a maximum steel content is also specified. Thus the maximum area of tension reinforcement in a beam should not exceed 4 per cent of the gross cross-sectional area of the concrete.

The area needed should generally be provided by not less than two bars and not more than eight bars. If necessary bars may be in groups of two, three or four, in contact. For the purpose of design such groups should be considered as a single bar of equivalent area. In addition the size of main bars used should normally not be less than 16 mm diameter.

The areas of round bar reinforcement are given in Table 3.8.

min R16 or T16s

Table 3.8 Areas of round bar reinforcement (mm²)

Diameter (mm)	Mass (kg/m)	Number of bars									
		1	2	3	4	5	6	7	8	9	10
6	0.222	28	57	85	113	142	170	198	226	255	283
8	0.395	50	101	151	201	252	302	352	402	453	502
10	0.617	79	157	236	314	393	471	550	628	707	785
12	0.888	113	226	339	452	565	678	791	904	1 017	1 130
16	1.58	201	402	603	804	1005	1206	1407	1 608	1 809	2 010
20	2.47	314	628	942	1256	1570	1884	2198	2 512	2 826	3 140
25	3.86	491	983	1474	1966	2457	2948	3439	3 932	4 423	4 915
32	6.31	804	1608	2412	3216	4020	4824	5628	6 432	7 236	8 040
40	9.87	1257	2513	3770	5027	6283	7540	8796	10 053	11 310	12 566

3.9.5 Minimum spacing of reinforcement

During concreting the aggregate must be allowed to move between the bars in order to achieve adequate compaction. For this reason BS 8110 Part 1 recommends a minimum bar spacing of 5 mm greater than the maximum coarse aggregate size h_{agg}. That is,

$$\text{Minimum distance between bars or group of bars} = h_{agg} + 5\,\text{mm}$$

When the diameter of the main bar or the equivalent diameter of the group is greater than $h_{agg} + 5\,\text{mm}$, the minimum spacing should not be less than the bar diameter or the equivalent diameter of the group.

A further consideration is the use of immersion type (poker) vibrators for compaction of the concrete. These are commonly 40 mm diameter, so that the space between bars to accommodate their use would have to be at least 50 mm.

3.9.6 Maximum spacing of reinforcement

When the limitation of crack widths to 0.3 mm is acceptable and the cover to reinforcement does not exceed 50 mm, the maximum bar spacing rules given in BS 8110 Part 1 may be adopted. Hence for the tension steel in

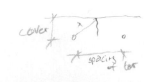

simply supported singly reinforced beams, the maximum clear distance between adjacent bars or groups would be as follows:

(a) When f_y is $250 \, \text{N/mm}^2$, clear distance $= 300 \, \text{mm}$
(b) When f_y is $460 \, \text{N/mm}^2$, clear distance $= 160 \, \text{mm}$.

3.9.7   Bending ULS

We know from the theory of bending that when bending is induced in a rectangular beam the material fibres above the neutral axis are subjected to compressive stresses and those below to tensile stresses. Concrete has excellent qualities for resisting compression. However, its resistance to tension is so poor that it is ignored. Instead, steel reinforcement is introduced to resist the tension.

On this basis simply supported rectangular beams are designed so that the concrete above the neutral axis is capable of resisting the induced compression, and tensile reinforcement capable of resisting the induced tension is introduced below the neutral axis. This reinforcement is positioned near the bottom of the beam where it will be most effective. Concrete beams designed in this way are described as singly reinforced.

For any beam to be adequate in bending, its internal moment of resistance must not be less than the externally applied bending moment. Therefore the design ultimate resistance moment M of a concrete beam must be greater than or at least equal to the ultimate bending moment M_u:

$$M \geqslant M_u$$

The ultimate bending moment is calculated in the normal manner but using the ultimate design loads. The design ultimate resistance moment of the beam can be obtained by reference to BS 8110 in any one of the following three ways:

(a) Using formulae derived from the short term stress/strain curves given in BS 8110 Part 1 Section 2.
(b) Using the design charts given in BS 8110 Part 3, which are based on the aforementioned stress/strain curves.
(c) Using the design formulae for rectangular beams given in BS 8110 Part 1, which are based on a simplified concrete stress block.

The singly reinforced rectangular beam examples included in this manual are based on method (c), for which the beam cross-section and stress diagram are shown in Figure 3.3. In one of the examples the result obtained using the formulae will be compared with that using the BS 8110 Part 3 design charts. For this purpose Chart 2 is reproduced here as Figure 3.4.

By considering the stress diagram shown in Figure 3.3 in conjunction with the theory of bending behaviour, the simplified design equations for

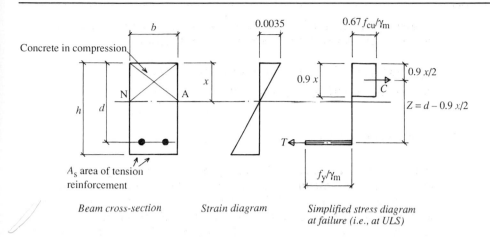

Figure 3.3 *Singly reinforced concrete beam diagrams*

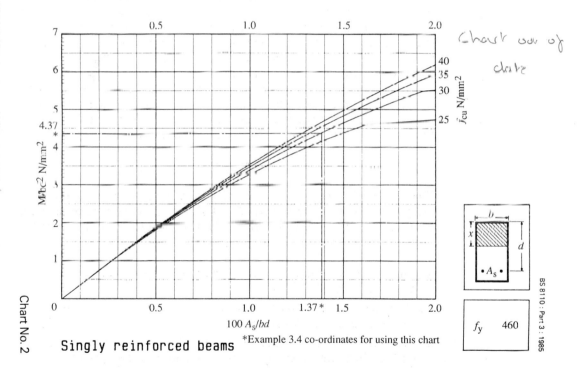

Figure 3.4 *BS 8110 Part 3 Chart 2 for singly reinforced beam design*

rectangular beams given in BS 8110 Part 1 can be derived. The relevant formulae for simply supported singly reinforced beams are as follows:

$$K = \frac{M}{bd^2 f_{cu}}$$

where for singly reinforced beams $K \leqslant K' = 0.156$,

$$z = d[0.5 + \sqrt{(0.25 - K/0.9)}] \not> 0.95d$$

$$x = \frac{d - z}{0.45}$$

$$A_s = \frac{M}{0.87 f_y z}$$

where

A_s area of tension reinforcement

b compression width of section

d effective depth of tension reinforcement

f_{cu} characteristic strength of concrete

f_y characteristic strength of steel

M design ultimate resistance moment: equals design ultimate moment M_u

x depth to neutral axis

z lever arm

It should be noted that the steel and concrete stresses used in these equations mean that the ultimate bending moment value M_u is in units N mm.

The manner in which a singly reinforced concrete beam fails in bending is influenced by the amount of reinforcement present in the section. If it is under-reinforced the tension steel will reach its yield stress before the concrete fails in compression. This would give ample warning of failure since excessive deflection would develop as failure approached. If it is over-reinforced then the concrete would fail prior to the tension steel reaching its yield stress. Such failure would occur suddenly without any significant deflection taking place.

To avoid sudden failure it is therefore important to ensure that the tension steel reaches its yield stress before the concrete fails in compression. Tests on beams have shown that the steel yields before the concrete crushes when the depth x to the NA does not exceed $0.5d$. By limiting K' to 0.156, BS 8110 implies that the NA depth does not exceed $0.5d$ and hence that the steel in tension will reach its ultimate stress before the concrete fails in compression.

If the value of K for a particular beam was found to be greater than the K' limit of 0.156 it would indicate that the concrete above the NA was overstressed in compression. Therefore either the beam would have to be increased in size, or compressive reinforcement would have to be introduced above the NA to assist the concrete. Design formulae and charts are given in BS 8110 for such beams, which are described as doubly reinforced.

When concrete roof or floor slabs are cast monolithically with the supporting beams, T or L beams are created as illustrated in Figure 3.5. Guidance is also given in BS 8110 for the design of these beams, which are collectively described as flanged beams.

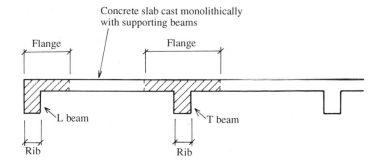

Figure 3.5 *Flanged beams*

3.9.8 Cracking SLS

Crack widths need to be controlled for appearance and to avoid corrosion of the reinforcement.

The cracking serviceability limit state will generally be satisfied by compliance with detailing rules given in BS 8110 Part 1. These relate to minimum reinforcement areas and bar spacing limits which for beams have already been stated in Sections 3.9.4 and 3.9.6. They ensure that crack widths will not exceed 0.3 mm.

Where it is necessary to limit crack widths to particular values less than 0.3 mm, perhaps for water tightness, then reference should be made to the guidance given in BS 8110 Part 2.

3.9.9 Deflection SLS

Reinforced concrete beams should be made sufficiently stiff that excessive deflections, which would impair the efficiency or appearance of the structure, will not occur. The degree of deflection allowed should be commensurate with the capacity of movement of any services, finishes, partitions, glazing, cladding and so on that the member may support or influence.

In all normal situations the deflection of beams will be satisfactory if the basic span to effective depth ratios given in BS 8110 Part 1 Table 3.10, reproduced here as Table 3.9, are not exceeded.

Table 3.9   Basic span to effective depth ratios for rectangular or flanged beams
(BS 8110 Part 1 1985 Table 3.10)

Support conditions	Rectangular sections	Flanged beams with $b_w/b \leqslant 0.3$
Cantilever	7	5.6
Simply supported	20	16.0
Continuous	26	20.8

It should be understood that the span to depth ratios given in Table 3.9 are based on the following:

(a) The span does not exceed 10 m.

(b) The total deflection is not greater than span/250, and the deflection after application of finishes and erection of partitions (that is, due to imposed loads) is limited to 20 mm or span/500.

(c) The loading pattern is uniformly distributed.

If for any reason it is necessary to change any of these parameters, the span to depth ratio must be adjusted accordingly:

(a) If the span exceeds 10 m,

$$\text{Revised ratio} = \text{basic ratio} \times 10/\text{span}$$

(b) If the total deflection must not exceed span/β,

$$\text{Revised ratio} = \text{basic ratio} \times 250/\beta$$

If deflection after the application of finishes must be less than the 20 mm limit,

$$\text{Revised ratio} = \text{basic ratio} \times \alpha/20$$

where α is the revised limit.

(c) The basic ratios have been derived with a uniformly distributed loading pattern coefficient of 0.104 included. Therefore for other loading patterns the basic ratios would have to be adjusted in proportion to the coefficient values. Reference should be made to BS 8110 Part 2 Table 3.1 for the coefficients for other loading patterns.

Deflection of beams is also influenced by the amount of tension steel present in the beam. This is expressed in terms of the design service stress in the steel and the ratio M/bd^2. To allow for this the basic span to effective depth ratio is modified by a factor from BS 8110 Part 1 Table 3.11, reproduced here as Table 3.10.

To determine the design service stress in the steel the following expression, from the second footnote to Table 3.10, is used:

$$f_s = \frac{2}{3} f_y \frac{A_{s,req}}{A_{s,prov}} \times \frac{1}{\beta_b}$$

where

$A_{s,prov}$ area of tension reinforcement provided at mid-span or at the support of a cantilever

$A_{s,req}$ area of tension reinforcement required

f_s estimated design service stress in the tension reinforcement

f_y characteristic strength of steel

β_b ratio to take account of any bending moment redistribution that has taken place

Table 3.10   Modification factor for tension reinforcement (BS 8110 Part 1 1985 Table 3.11)

Service stress	M/bd^2								
	0.50	0.75	1.00	1.50	2.00	3.00	4.00	5.00	6.00
100	2.00	2.00	2.00	1.86	1.63	1.36	1.19	1.08	1.01
150	2.00	2.00	1.98	1.69	1.49	1.25	1.11	1.01	0.94
156 ($f_y = 250$)	2.00	2.00	1.96	1.66	1.47	1.24	1.10	1.00	0.94
200	2.00	1.95	1.76	1.51	1.35	1.14	1.02	0.94	0.88
250	1.90	1.70	1.55	1.34	1.20	1.04	0.94	0.87	0.82
288 ($f_y = 460$)	1.68	1.50	1.38	1.21	1.09	0.95	0.87	0.82	0.78
300	1.60	1.44	1.33	1.16	1.06	0.93	0.85	0.80	0.76

Note 1: The values in the table derive from the equation

$$\text{Modification factor} = 0.55 + \frac{477 - f_s}{120(0.9 + M/bd^2)} \leqslant 2.0$$

where M is the design ultimate moment at the centre of the span or, for a cantilever, at the support.

Note 2: The design service stress in the tension reinforcement in a member may be estimated from the equation

$$f_s = \frac{5 f_y A_{s,req}}{8 A_{s,prov}} \times \frac{1}{\beta_h}$$

Note 3: For a continuous beam, if the percentage of redistribution is not known but the design ultimate moment at mid-span is obviously the same as or greater than the elastic ultimate moment, the stress f_s in this table may be taken as $5/8 f_y$.

The ratio β_b does not apply to the simply supported beams dealt with in this manual. Hence the expression for simply supported beams becomes

$$f_s = \frac{5}{8} f_y \frac{A_{s,req}}{A_{s,prov}}$$

Using the tables, the minimum effective depth d for a singly reinforced beam to satisfy deflection requirements may be written as follows:

$$\text{Minimum effective depth } d = \frac{\text{span}}{\text{Table 3.9 factor} \times \text{Table 3.10 factor}}$$

Before proceeding to examine the effect of shear on concrete beams, let us look at some examples on the bending and deflection requirements.

Example 3.4

The singly reinforced concrete beam shown in Figure 3.6 is required to resist an ultimate moment of 550 kN m. If the beam is composed of grade 30 concrete and high yield (HY) reinforcement, check the section size and determine the area of steel required.

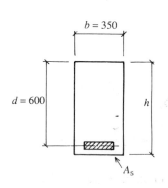

Figure 3.6   *Beam cross-section*

In this example the breadth and effective depth dimensions which are given can be checked by using the BS 8110 simplified stress block formulae.

Grade 30 $f_{cu} = 30 \, \text{N/mm}^2$

HY steel reinforcement $f_y = 460 \, \text{N/mm}^2$

Ultimate bending moment $M_u = 550 \, \text{kNm} = 550 \times 10^6 \, \text{N mm}$

Using BS 8110 formulae:

$$K = \frac{M}{bd^2 f_{cu}} = \frac{550 \times 10^6}{350 \times 600^2 \times 30} = 0.146 < K' = 0.156$$

Therefore compression reinforcement is not necessary and the section size is adequate for a singly reinforced beam.

The area of tension reinforcement needed may now be determined. The lever arm is given by

$$z = d[0.5 + \sqrt{(0.25 - K/0.9)}] = d[0.5 + \sqrt{(0.25 - 0.146/0.9)}]$$
$$= 0.796d < 0.95d$$

This is satisfactory. Therefore

$$A_s \text{ required} = \frac{M}{0.87 f_y z} = \frac{550 \times 10^6}{0.87 \times 460 \times 0.796 \times 600} = 2878 \, \text{mm}^2$$

This area can be compared with the reinforcement areas given in Table 3.8 to enable suitable bars to be selected:

Provide six 25 mm diameter HY bars in two layers ($A_s = 2948 \, \text{mm}^2$).

The overall beam depth, as shown in Figure 3.7, can now be determined by reference to the dimensional requirements for beams and by assuming that the

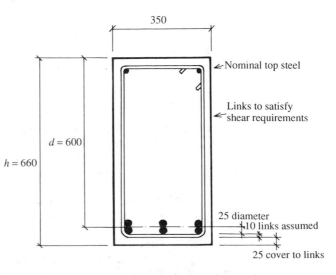

Figure 3.7 *Finalized beam cross-section*

beam will be in a situation of mild exposure and be provided with 10 mm diameter links:

Effective depth d	600
Diameter of lower bars	25
Diameter of links assumed	10
Cover, mild exposure assumed	25
Overall depth h	660 mm

Provide a 660 mm × 350 mm grade 30 concrete beam.

The percentage reinforcement area provided can now be compared with the requirements of BS 8110:

$$\text{Steel content} = \frac{2948}{660 \times 350} \times 100 = 1.28 \text{ per cent}$$

This is greater than the minimum area of 0.13 per cent required for high yield steel and less than the maximum area of 4 per cent required for all steels. The beam is therefore adequate for the bending ULS.

For comparison let us now look at the use of the BS 8110 Part 3 design charts for this example. Chart 2, shown in Figure 3.4, relates to singly reinforced beams containing tension reinforcement with a yield stress of 460 N/mm². We have

$$\frac{M}{bd^2} = \frac{550 \times 10^6}{350 \times 600^2} = 4.37$$

From the chart this gives $100A_s/bd = 1.37$ for an f_{cu} of 30 N/mm². By rearranging this expression the area of tensile steel required can be determined:

$$A_s \text{ required} = 1.37\frac{bd}{100} = \frac{1.37 \times 350 \times 600}{100} = 2877 \text{ mm}^2$$

This is the area previously obtained using the design formulae, and we would therefore provide the same bars.

The remaining calculations needed to complete the design are exactly as those made previously.

Example 3.5

A reinforced concrete beam is required to transmit an ultimate bending moment of 140 kN m, inclusive of its own weight. Using the simplified stress block formulae given in BS 8110 Part 1, determine the depth of beam required and the amount of steel needed in a 250 mm wide beam for the following combinations:

Grade 30 concrete with mild steel reinforcement

Grade 35 concrete with high yield reinforcement.

As can be seen from the beam cross-section shown in Figure 3.8, the breadth b is known but the effective depth d, the overall depth h and the area of tensile reinforcement A_s are not. The formulae must therefore be used in conjunction with the ultimate bending moment to determine these values.

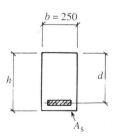

Figure 3.8 *Beam cross-section*

Grade 30 concrete, MS reinforcement

Grade 30 concrete $f_{cu} = 30 \, \text{N/mm}^2$

MS reinforcement $f_y = 250 \, \text{N/mm}^2$

Ultimate bending moment $M_u = 140 \, \text{kN m} = 140 \times 10^6 \, \text{N mm}$

Consider the BS 8110 formulae. First $K = M/bd^2 f_{cu}$, from which an expression for d may be derived. In addition, for singly reinforced beams $K \leqslant K' = 0.156$. Hence

$$d \text{ required} = \sqrt{\left(\frac{M}{Kbf_{cu}}\right)} = \sqrt{\left(\frac{M}{0.156bf_{cu}}\right)}$$

Also, if $K = K' = 0.156$,

$$z = d[0.5 + \sqrt{(0.25 - K/0.9)}] = d[0.5 + \sqrt{(0.25 - 0.156/0.9)}] = 0.777d \ngtr 0.95d$$

Hence

$$A_s = \frac{M}{0.87f_y z} = \frac{M}{0.87f_y 0.777d}$$

Apply these expressions to the beam in question:

$$d \text{ required} = \sqrt{\left(\frac{M}{0.156bf_{cu}}\right)} = \sqrt{\left(\frac{140 \times 10^6}{0.156 \times 250 \times 30}\right)} = 345.92 \, \text{mm:} \quad \text{use } 350 \, \text{mm}$$

$$A_s \text{ required} = \frac{M}{0.87f_y 0.777d} = \frac{140 \times 10^6}{0.87 \times 250 \times 0.777 \times 350} = 2367 \, \text{mm}^2$$

Provide three 25 mm diameter and three 20 mm diameter MS bars in two layers ($A_s = 1474 + 942 = 2416 \, \text{mm}^2$).

It should be appreciated that since a larger effective depth d than that required has been adopted, strictly speaking a revised value of K and z should be calculated. However, this would have no practical effect on the solution as the same size and number of bars would still be provided.

To determine the beam depth (see Figure 3.9):

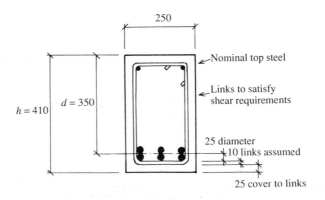

Figure 3.9 *Beam cross-section: grade 30 concrete, MS reinforcement*

Effective depth d	350
Diameter of lower bars	25
Diameter of links assumed	10
Cover, mild exposure assumed	25
Overall beam depth h	410 mm

Provide a 410 mm × 250 mm grade 30 concrete beam. Check percentage steel content:

$$\text{Steel content} = \frac{2416}{410 \times 250} \times 100 = 2.35 > 0.24 < 4 \text{ per cent}$$

The beam is adequate for the bending ULS.

Grade 35 concrete, HY reinforcement
Grade 35 concrete $f_{cu} = 35 \text{ N/mm}^2$
HY reinforcement $f_y = 460 \text{ N/mm}^2$
$M_u = 140 \text{ kN m} = 140 \times 10^6 \text{ N mm}$

$$d \text{ required} = \sqrt{\left(\frac{M}{0.156 b f_{cu}}\right)} = \sqrt{\left(\frac{140 \times 10^6}{0.156 \times 250 \times 35}\right)} = 320.25 \text{ mm:} \quad \text{use 325 mm}$$

$$A_s \text{ required} = \frac{M}{0.87 f_y 0.777 d} = \frac{140 \times 10^6}{0.87 \times 460 \times 0.777 \times 325} = 1385 \text{ mm}^2$$

Provide three 25 mm diameter HY bars ($A_s = 1474 \text{ mm}^2$).

Determine the beam depth (see Figure 3.10):

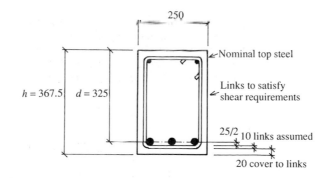

Figure 3.10 *Beam cross-section: grade 35 concrete, HY reinforcement*

Effective depth d	325
Main bar diameter/2	12.5
Diameter of links assumed	10
Cover, mild exposure assumed	20
Overall beam depth h	367.5 mm

Provide a 370 mm × 250 mm grade 35 concrete beam. Check percentage steel content:

$$\text{Steel content} = \frac{1474}{370 \times 250} \times 100 = 1.59 > 0.13 < 4 \text{ per cent}$$

The beam is adequate for the bending ULS.

Example 3.6

A reinforced concrete beam with an effective span of 7 m is 500 mm deep overall by 250 mm wide. It supports a characteristic imposed load of 9 kN per metre run and a characteristic dead load of 11 kN per metre run in addition to the load due to the beam self-weight, which may be taken as 24 kN/m^3. Using the simplified stress block formulae given in BS 8110 Part 1, check that the beam depth is adequate. Choose suitable tension reinforcement if the steel has a yield stress of 460 N/mm^2 and the concrete is grade 30.

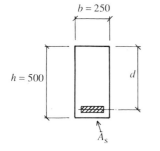

Figure 3.11 *Beam cross-section*

The beam cross-section is shown in Figure 3.11.

First let us ensure that the slenderness limits for beams are satisfied to avoid lateral buckling. Maximum distance between lateral restraints is the lesser of

$$60b_c = 60 \times 250 = 15\,000 \text{ mm} = 15 \text{ m}$$

$$250\frac{b_c^2}{d} = \frac{250 \times 250^2}{500} = 31\,250 \text{ mm} = 31.25 \text{ m}$$

These are both greater than the 7 m effective span of the beam and therefore the slenderness limits are satisfied.

Now let us calculate the ultimate design load and the ultimate bending moment:

Characteristic imposed Q_k UDL = 9 kN/m = 9 × 7 = 63 kN

Characteristic dead G_k UDL = 11 kN/m = 11 × 7 = 77 kN

Characteristic own weight G_k UDL = 24 × 7 × 0.5 × 0.25 = 21 kN

ULS imposed	UDL = $(\gamma_f Q_k)$	= 1.6 × 63 = 100.8 kN
ULS dead	UDL = $(\gamma_f G_k)$	= 1.4 × 77 = 107.8 kN
ULS self-weight UDL = $(\gamma_f SW)$	= 1.4 × 21 =	29.4 kN
ULS total	UDL =	238 kN

The beam load diagram is shown in Figure 3.12. Thus

$$M_u = \frac{WL}{8} = \frac{238 \times 7}{8} = 208.25 \text{ kN m} = 208.25 \times 10^6 \text{ N mm}$$

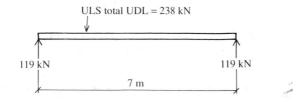

Figure 3.12 *Beam load diagram*

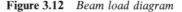

In order to use the expression for K in this example an effective depth d will have to be assumed. If it is assumed that a single layer of 25 mm diameter main bars will be adopted with 10 mm diameter links and 25 mm cover, the resulting effective depth would be as follows:

Assumed effective depth $d = 500 - \dfrac{25}{2} - 10 - 25 = 452.5$ mm

Then

$$K = \frac{M}{bd^2 f_{cu}} = \frac{208.25 \times 10^6}{250 \times 452.5^2 \times 30} = 0.136 < 0.156$$

This is satisfactory. Then the lever arm is given by

$$z = d[0.5 + \sqrt{(0.25 - K/0.9)}] = d[0.5 + \sqrt{(0.25 - 0.136/0.9)}] = 0.814d$$

Hence

$$A_s = \frac{M}{0.87 f_y z} = \frac{208.25 \times 25 \times 10^6}{0.87 \times 460 \times 0.814 \times 452.5} = 1413 \text{ mm}^2 \qquad P81$$

Provide three 25 mm diameter HY bars ($A_s = 1474$ mm^2). Since 25 mm diameter main bars have been adopted, the effective depth assumed was correct and therefore the overall depth of 500 mm is satisfactory. Check percentage steel content:

$$\text{Steel content} = \frac{1474}{500 \times 250} \times 100 = 1.18 > 0.13 < 4 \text{ per cent}$$

The beam is adequate in bending.

The deflection SLS can be checked by reference to the recommended span to depth ratios given in BS 8110. The basic span to effective depth ratio (from Table 3.9) is 20. This must be modified by the factor for the amount of tension reinforcement in the beam, obtained from Table 3.10. So

$$\frac{M}{bd^2} = \frac{208.25 \times 10^6}{250 \times 452.5^2} = 4.07$$

From the expression given in the second footnote to the table:

$$f_s = \frac{5}{8} f_y \frac{A_{s,req}}{A_{s,prov}} = \frac{5}{8} \times 460 \times \frac{1413}{1474} = 276$$

The modification factor may be obtained by interpolation from Table 3.10 or by using the formula given at the foot of the table:

$$\text{Modification factor} = 0.55 + \frac{477 - f_s}{120(0.9 + M/bd^2)}$$

$$= 0.55 + \frac{477 - 276}{120(0.9 + 4.07)} = 0.887 \leqslant 2$$

Therefore the allowable span to effective depth ratio modified for tension reinforcement is $20 \times 0.887 = 17.74$. Finally,

$$\text{Actual span to effective depth ratio} = \frac{7000}{452.5} = 15.47 < 17.74 \quad \text{less ok}$$

Hence the beam is adequate in deflection.

3.9.10   Shear ULS

The effect of shear at the ULS must be examined for all concrete beams. Except in members of minor structural importance, such as lintels, some form of shear reinforcement should be introduced. Such reinforcement may consist of either vertical links alone or vertical links combined with bent-up bars.

Shear failure in concrete beams is of a complex nature and can occur in several ways. A typical failure mode, for a simply supported beam, is illustrated in Figure 3.13. The manner in which links and bent-up bars assist in resisting shear is also shown.

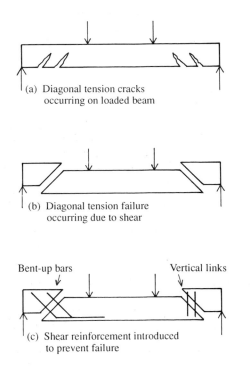

Figure 3.13   *Typical failure mode due to shear for a simply supported beam*

The procedure for checking the shear resistance of concrete beams is carried out in the following manner. First, calculate the design shear stress occurring from

$$v = \frac{V}{b_v d}$$

where

v design shear stress occurring at cross-section being considered

V design shear force due to ultimate loads

b_v breadth of section

d effective depth

To avoid diagonal compression failure of the concrete when shear reinforcement is present, v must never exceed the lesser of $0.8\sqrt{f_{cu}}$ or 5 N/mm². If the calculated value of v did exceed these limits then the size of beam would need to be increased.

The next step is to determine the form and area of the shear reinforcement needed by comparing the calculated value of v with the guidance given in BS 8110 Part 1 Table 3.8, and the design shear stress capacity of concrete v_c given in BS 8110 Table 3.9. These tables are reproduced here as Tables 3.11 and 3.12 respectively.

Table 3.11 Form and area of shear reinforcement in beams (BS 8110 Part 1 1985 Table 3.8)

Value of v (N/mm²)	Form of shear reinforcement to be provided	Area of shear reinforcement to be provided
Less than $0.5v_c$ throughout the beam	See note 1	
$0.5v_c < v < (v_c + 0.4)$	Minimum links for whole length of beam	$A_{sv} \geqslant 0.4b_v s_v/0.87f_{yv}$ (see Note 2)
$(v_c + 0.4) < v < 0.8\sqrt{f_{cu}}$ or 5 N/mm²	Links or links combined with bent-up bars. Not more than 50% of the shear resistance provided by the steel may be in the form of bent-up bars (see Note 3)	Where links only provided: $A_{sv} \geqslant b_v s_v(v - v_c)/0.87f_{yv}$ Where links and bent-up bars provided: see clause 3.4.5.6 of BS 8110

Note 1: While minimum links should be provided in all beams of structural importance, it will be satisfactory to omit them in members of minor structural importance such as lintels or where the maximum design shear stress is less than $0.5v_c$.

Note 2: Minimum links provide a design shear resistance of 0.4 N/mm².

Note 3: See clause 3.4.5.5 of BS 8110 for guidance on spacing of links and bent-up bars.

Table 3.12 Values of design concrete shear stress v_c (N/mm²) (BS 8110 Part 1 1985 Table 3.9)

$100A_s/b_v d$	Effective depth (mm)							
	125	150	175	200	225	250	300	$\geqslant 400$
$\leqslant 0.15$	0.45	0.43	0.41	0.40	0.39	0.38	0.36	0.34
0.25	0.53	0.51	0.49	0.47	0.46	0.45	0.43	0.40
0.50	0.67	0.64	0.62	0.60	0.58	0.56	0.54	0.50
0.75	0.77	0.73	0.71	0.68	0.66	0.65	0.62	0.57
1.00	0.84	0.81	0.78	0.75	0.73	0.71	0.68	0.63
1.50	0.97	0.92	0.89	0.86	0.83	0.81	0.78	0.72
2.00	1.06	1.02	0.98	0.95	0.92	0.89	0.86	0.80
$\geqslant 3.00$	1.22	1.16	1.12	1.08	1.05	1.02	0.98	0.91

Note 1: Allowance has been made in these figures for a γ_m of 1.25.

Note 2: The values in the table are derived from the expression

$$0.79[100A_s/(b_v d)]^{1/3}(400/d)^{1/4}/\gamma_m$$

where $100A_s/b_v d$ should not be taken as greater than 3, and $400/d$ should not be taken as less than 1. For characteristic concrete strengths greater than 25 N/mm², the values in the table may be multiplied by $(f_{cu}/25)^{1/3}$. The value of f_{cu} should not be taken as greater than 40.

Two points need to be appreciated with respect to the use of Table 3.12. First, the tabulated values of v_c only apply to grade 25 concrete. For higher characteristic strengths up to a limiting f_{cu} of 40 N/mm² , the values may be increased by multiplying them by $(f_{cu}/25)^{1/3}$.

Second, the percentage of main tensile reinforcement in the member under consideration should not be taken, for the purpose of the shear calculations, as greater than 3 per cent. Nor, again for the purpose of the shear calculations, should its effective depth be taken as greater than 400 mm.

The guidance given in Table 3.11 will establish whether and in what form shear reinforcement is required, according to three values of the shear stress v:

(a) $v < 0.5v_c$

Theoretically no shear reinforcement is necessary throughout the length of the beam. However, with the exception of simple lintels, nominal reinforcement in the form of minimum links should be provided in all beams of structural importance.

(b) $0.5v_c < v < (v_c + 0.4)$

Only minimum links are required.

(c) $(v_c + 0.4) < v < 0.8\sqrt{f_{cu}}$ or 5 N/mm²

Designed links, or a combination of designed links and bent-up bars, are necessary.

The procedures for (b) and (c) are described in the following sections.

In certain circumstances, near to supports, advantage may be taken of an enhanced shear strength, for which guidance is given in clause 3.4.5.8 of BS 8110 Part 1.

Minimum links

When minimum links are to be provided as shown in Figure 3.14, their area should be determined from the following expression:

$$A_{sv} \geqslant \frac{0.4b_v s_v}{0.87f_{yv}}$$

where

A_{sv} total cross-section of links at the neutral axis, at a section
b_v breadth of section
f_{yv} characteristic strength of links (that is 250 N/mm² or 460 N/mm²)
s_v spacing of links along the member

BS 8110 states that the spacing of links should not exceed 0.75d. Hence,

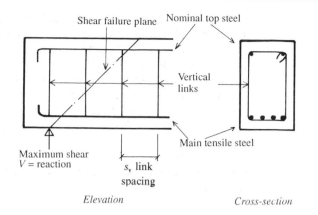

Figure 3.14 *Shear reinforcement in the form of vertical links*

as a trial, this limit may be substituted in the area formula as follows:

$$A_{sv} \geqslant \frac{0.4b_v 0.75d}{0.87f_{yv}}$$

Should the resulting area prove impractical the link spacing may of course be reduced.

Designed links

When shear reinforcement greater than minimum links is necessary, it may be provided either as designed links alone or as designed links combined with bent-up bars. In both instances, it must be capable of resisting the difference between the applied design shear stress v and the design shear stress capacity of the concrete v_c.

Where designed links alone are to be provided, their area should be determined from the following expression:

$$A_{sv} \geqslant \frac{b_v s_v (v - v_c)}{0.87f_{yv}}$$

The symbols and maximum spacing are as for minimum links.

Designed links and bent-up bars

Where shear reinforcement needs to be provided in the form of designed links combined with bent-up bars, the total shear resistance capacity will be the summation of the individual values for each system. In this context the contribution made by the bent-up bars should not be taken as more than 50 per cent of the total shear resistance. The shear resistance of the designed links may be determined from the information given above, whilst that of the bent-up bars is discussed in the following.

Bent-up bars, as their name implies, are main tension bars that are bent up at an angle from the bottom of the beam as shown in Figure 3.15. Such bars cannot be bent up unless they are no longer required to resist the bending moment in the tension zone. This is only likely to be the case near to a support where the bending moment is reducing and hence fewer bars are needed in tension. Their design shear resistance is based upon the assumption that they act as tension members in an imaginary truss system, whilst the concrete forms the compression members as shown in Figure 3.16. The truss should be arranged so that α and β are both greater than or equal to 45°, giving a maximum value s_t of $1.5d$.

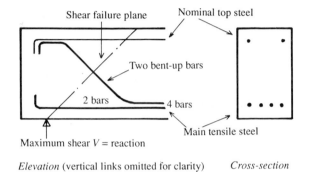

Elevation (vertical links omitted for clarity)   *Cross-section*

Figure 3.15 *Shear reinforcement in the form of bent-up bars*

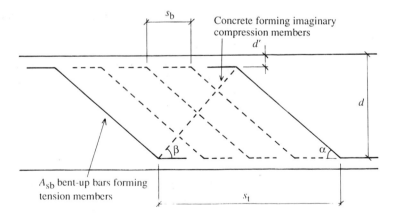

Figure 3.16 *Imaginary truss system of bent-up bars*

The shear resistance of bent-up bars in such systems should be calculated from the following expression:

$$V_b = A_{sb}(0.87f_{yv})(\cos \alpha + \sin \alpha \cot \beta)\frac{d - d'}{s_b}$$

The symbols are indicated in Figure 3.16 or have been referred to earlier.

Now let us look at some examples on the design of shear reinforcement.

Example 3.7

Determine the form of shear reinforcement to be provided in a 150 mm wide concrete lintel that has an effective depth of 200 mm and supports an ultimate UDL of 18.6 kN. The lintel is cast from grade 25 concrete and contains tensile reinforcement with an area of 226 mm².

Ultimate design shear force at support $V = \dfrac{\text{ultimate UDL}}{2} = \dfrac{18.6}{2} = 9.3 \text{ kN}$

Maximum design shear stress occurring $v = \dfrac{V}{b_v d} = \dfrac{9.3 \times 10^3}{150 \times 200} = 0.31 \text{ N/mm}^2$

Hence

$$v = 0.31 \text{ N/mm}^2 < 0.8\sqrt{f_{cu}} = 4 \text{ N/mm}^2 < 5 \text{ N/mm}^2$$

Therefore the beam size is satisfactory.
Now

$$\frac{100 A_s}{b_v d} = \frac{100 \times 226}{150 \times 200} = 0.75$$

Thus the design concrete shear stress (from Table 3.12) $v_c = 0.68 \text{ N/mm}^2$, and $0.5 v_c = 0.5 \times 0.68 = 0.34 \text{ N/mm}^2$. Hence $v = 0.31 \text{ N/mm}^2$ is less than $0.5 v_c = 0.34 \text{ N/mm}^2$. Therefore, by reference to Table 3.11 note 1, it would be satisfactory to omit links from this member since it is a lintel.

Example 3.8

A reinforced concrete beam supporting an ultimate UDL of 240 kN is 250 mm wide with an effective depth of 500 mm. If the concrete is grade 30 and the area of tensile steel provided is 1256 mm², determine the form and size of shear reinforcement required.

Ultimate design shear force at support $V = \dfrac{\text{ultimate UDL}}{2} = \dfrac{240}{2} = 120 \text{ kN}$

Maximum design shear stress occurring $v = \dfrac{V}{b_v d} = \dfrac{120 \times 10^3}{250 \times 500} = 0.96 \text{ N/mm}^2$

Hence

$$v = 0.96 \text{ N/mm}^2 < 0.8\sqrt{f_{cu}} = 4.38 \text{ N/mm}^2 < 5 \text{ N/mm}^2$$

Therefore the beam size is satisfactory.
Now

$$\frac{100 A_s}{b_v d} = \frac{100 \times 1256}{250 \times 500} = 1$$

Thus the design concrete shear stress (from Table 3.12) $v_c = 0.63 \, \text{N/mm}^2$. This is based on the maximum effective depth limit of 400 mm and is for grade 25 concrete. As the concrete in this example is grade 30, the value of v_c may be multiplied by a coefficient:

$$\text{Coefficient} = (f_{cu}/25)^{1/3} = (30/25)^{1/3} = 1.062$$

Thus

$$\text{Grade 30 } v_c = 0.63 \times 1.062 = 0.669 \, \text{N/mm}^2$$
$$0.5v_c = 0.5 \times 0.669 = 0.335 \, \text{N/mm}^2$$

Also

$$(v_c + 0.4) = (0.669 + 0.4) = 1.069 \, \text{N/mm}^2$$

Hence $0.5v_c$ is less than v, which is less than $(v_c + 0.4)$. Therefore, by reference to Table 3.11, shear reinforcement in the form of minimum links should be provided for the whole length of the beam.

Some of the alternative ways of providing shear reinforcement in the form of links are illustrated in Figure 3.17. When considering the most suitable arrangement for the links the following points should be taken into account:

(a) The horizontal spacing should be such that no main tensile reinforcing bar should be further than 150 mm away from a vertical leg of the links.
(b) The horizontal spacing of the link legs across the section should not exceed the effective depth d.
(c) The horizontal spacing along the span should not exceed $0.75d$.

The area of the minimum links to be provided is determined from the relevant formula given in Table 3.11:

$$A_{sv} \geqq \frac{0.4b_v s_v}{0.87f_{yv}}$$

Since the spacing s_v must not exceed $0.75d$, this value may be substituted in the formula as a trial:

$$A_{sv} = \frac{0.4b_v 0.75d}{0.87f_{yv}}$$

Thus if mild steel links are to be provided with $f_{yv} = 250 \, \text{N/mm}^2$,

$$A_{sv} \text{ required} = \frac{0.4 \times 250 \times 0.75 \times 500}{0.87 \times 250} = 172.41 \, \text{mm}^2$$

In order to provide an area A_{sv} greater than 172.41 mm^2 it would be necessary to use 12 mm diameter links with an A_{sv} for two legs of 226 mm^2. If for practical reasons it was desired to use smaller links, either high yield links could be used or the centres could be reduced to suit smaller diameter mild steel links.

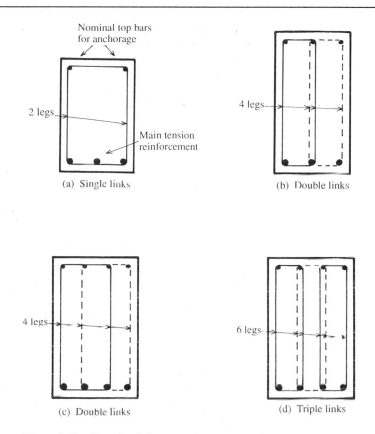

Figure 3.17 *Details of shear reinforcement in the form of links*

For high yield links the f_{yv} would become 460 N/mm², and

$$A_{sv} \text{ required} = \frac{0.4 \times 250 \times 0.75 \times 500}{0.87 \times 460} = 93.7 \text{ mm}^2$$

Provide 8 mm diameter HY links at 375 mm centres (A_{sv} two legs = 101 mm²).

Alternatively, if it were desired to use 8 mm diameter mild steel links the formula could be transposed to calculate the necessary centres s_v:

$$s_v = \frac{0.87 f_{yv} A_{sv}}{0.4 b_v} = \frac{0.87 \times 250 \times 101}{0.4 \times 250} = 220 \text{ mm} < 0.75d$$

Provide 8 mm diameter MS links at 200 mm centres.

If considered practical it is possible to provide double links or even triple links as shown in Figure 3.17. In this example it would not be practical to use triple links in a beam width of 250 mm, but double links could be provided. For mild steel double links, the total area of their four legs would have to be greater than the A_{sv} of 172.41 mm² previously calculated. Hence:

Provide 8 mm diameter MS double links at 375 mm centres (A_{sv} four legs = 201 mm²).

Example 3.9

A reinforced concrete beam 300 mm wide with an effective depth of 450 mm supports an ultimate UDL of 880 kN. Determine the form and size of shear reinforcement required if the main tensile reinforcement area is 2368 mm^2 and the concrete is grade 35.

Ultimate design shear force at support $V = \dfrac{880}{2} = 440 \, \text{kN}$

Maximum design shear stress occurring $v = \dfrac{V}{b_v d} = \dfrac{440 \times 10^3}{300 \times 450} = 3.26 \, \text{N/mm}^2$

Hence

$$v = 3.26 \, \text{N/mm}^2 < 0.8\sqrt{f_{cu}} = 4.73 \, \text{N/mm}^2 < 5 \, \text{N/mm}^2$$

Therefore the beam size is satisfactory.
  Now

$$\frac{100 A_s}{b_v d} = \frac{100 \times 2368}{300 \times 450} = 1.75$$

Thus the design concrete shear stress (from Table 3.12) $v_c = 0.76 \, \text{N/mm}^2$. This is based on the maximum effective depth of 400 mm and is for grade 25 concrete. For grade 35 concrete,

Coefficient $= (f_{cu}/25)^{1/3} = (35/25)^{1/3} = 1.119$

Thus

Grade 35 $v_c = 0.76 \times 1.119 = 0.85 \, \text{N/mm}^2$
  $(v_c + 0.4) = (0.85 + 0.4) = 1.25 \, \text{N/mm}^2$

Hence $(v_c + 0.4)$ is less than v which is less than $0.8\sqrt{f_{cu}}$. Therefore, by reference to Table 3.11, shear reinforcement in the form of either designed links alone or designed links combined with bent-up bars should be provided. Whilst 50 per cent of the shear resistance may be provided by bent-up bars they are not recommended as a practical choice and are usually avoided in favour of designed links alone.
  Designed links alone will therefore be adopted. Their area is calculated from the relevant formula given in Table 3.11:

$$A_{sv} \geqq \frac{b_v s_v (v - v_c)}{0.87 f_{yv}}$$

Since s_v must not exceed $0.75d$ this may again be substituted in the formula as a trial:

$$A_{sv} = \frac{b_v 0.75 d (v - v_c)}{0.87 f_{yv}}$$

Thus if mild steel links are to be provided with f_{yv} of 250 N/mm^2,

$$A_{sv} \text{ required} = \frac{300 \times 0.75 \times 450(3.26 - 0.85)}{0.87 \times 250} = 1121.89 \, \text{mm}^2$$

However, since four 20 mm diameter legs are needed, this is not practical in mild steel. High yield steel would not be much better: pro rata for HY steel,

$$A_{sv} \text{ required} = \frac{1121.89 \times 250}{460} = 609.72 \text{ mm}^2$$

Therefore try assuming a more practical diameter and determine the required centres by transposing the formula. Assuming 10 mm diameter MS double links, $A_{sv} = 314 \text{ mm}^2$. Thus

$$A_{sv} = \frac{b_v s_v (v - v_c)}{0.87 f_{yv}}$$

$$s_v = \frac{0.87 f_{yv} A_{sv}}{b_v(v - v_c)} = \frac{0.87 \times 250 \times 314}{300(3.26 - 0.85)} = 94 \text{ mm} < 0.75d$$

Provide 10 mm diameter MS double links at 90 mm centres.
Alternatively, assuming 10 mm diameter HY double links,

$$s_v = \frac{0.87 \times 460 \times 314}{300(3.26 - 0.85)} = 173.81 \text{ mm} < 0.75d$$

Provide 10 mm diameter HY double links at 170 mm centres.
It should be appreciated that it may be practical to increase the spacing of links towards mid-span as the shear force reduces.

3.9.11   Design summary for concrete beams

The design procedure for simply supported singly reinforced concrete beams may be summarized as follows:

(a) Calculate the ultimate loads, shear force and bending moment acting on the beam.
(b) Check the bending ULS by reference to the BS 8110 simplified stress block formulae. This will determine an adequate depth for the beam singly reinforced and the area of tension reinforcement required.
(c) Ensure that the cracking SLS is satisfied by compliance with the recommendations for minimum reinforcement content and bar spacing.
(d) Check the deflection SLS by reference to the recommended span to depth ratios.
(e) Check the shear ULS by providing the relevant link reinforcement in accordance with the guidance given in BS 8110.

3.10   Slabs

BS 8110 deals with suspended slabs as opposed to ground bearing slabs. For guidance on the design of the latter, reference should be made to other sources such as the literature published by the British Cement Association, formerly known as the Cement and Concrete Association.

Suspended slabs may be designed to span in either one or two directions depending on how they are supported at the edges.

In the context of BS 8110, slabs are classified into three groups:

Solid slabs   These, as the name implies, consist of solid concrete reinforced where necessary to resist tension (Figure 3.18).

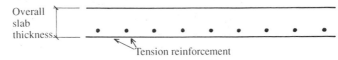

Figure 3.18   *Cross-section through a solid slab*

Ribbed slabs   For spans exceeding 4 m the self-weight of solid slabs can begin to affect their economy. In such circumstances consideration should be given to the use of ribbed slabs. These are formed in any one of the following ways:

(a) As a series of *in situ* concrete ribs cast between hollow or solid block formers which remain part of the completed slab (Figure 3.19).

(b) As a series of *in situ* concrete ribs cast monolithically with the concrete topping on removable forms (Figure 3.20).

(c) As an apparently solid slab but containing permanent formers to create voids within the cross-section (Figure 3.21).

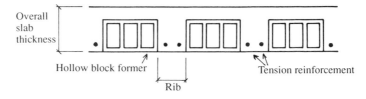

Figure 3.19   *Cross-section through a ribbed slab cast with integral hollow block formers*

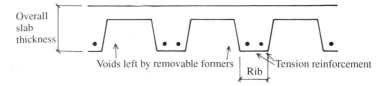

Figure 3.20   *Cross-section through a ribbed slab cast on removable formers*

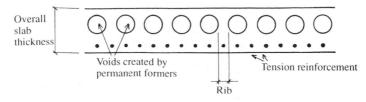

Figure 3.21   *Cross-section through a hollow slab cast with permanent void formers*

Flat slabs   The title of such slabs is descriptively something of a misnomer. It is intended to describe slabs which have been designed to act in conjunction with columns as a structural frame without the necessity for beams, and hence have a flat soffit (Figure 3.22). They can however have thickened sections where the soffit is dropped to form a stiffening band running between the columns (Figure 3.23). The top of the columns may also be enlarged locally by the formation of a column head to give support to the slab over a larger area (Figure 3.24). Flat slabs may be solid or may have recesses formed in the soffit to give a series of two-directional ribs, in which case they are often referred to as waffle or coffered slabs.

Figure 3.22   *Section through a flat slab*   **Figure 3.23**   *Section through a flat slab with drops*

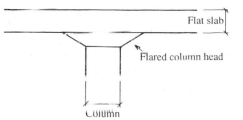

Figure 3.24   *Section through a flat slab with enlarged column heads*

The most commonly encountered suspended slabs are those used for the floors and roofs of buildings. However, sloping slabs are also used to form ramps, and concrete staircases are in fact a type of cranked slab.

For the purpose of this manual only the design of solid slabs spanning in one direction will be studied. Their design will be examined under the following headings, and where relevant a comparison will be made with the considerations for beams given in Section 3.9:

(a) Dimensional considerations
(b) Reinforcement areas
(c) Minimum spacing of reinforcement
(d) Maximum spacing of reinforcement

(e) Bending ULS
(f) Cracking SLS
(g) Deflection SLS
(h) Shear ULS.

Therefore let us consider how each of these influences the design of slabs.

3.10.1   Dimensional considerations

The two principal dimensional considerations for a one-way spanning slab are its width and its effective span.

For the purpose of design, reinforced concrete slabs spanning in one direction are considered as a series of simply supported beams having some convenient width, usually taken to be 1 m as shown in Figure 3.25. Their effective span or length is the same as that for beams given in Section 3.9.1.

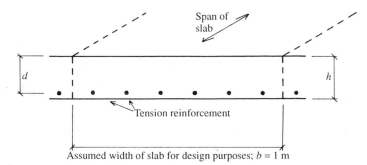

Figure 3.25   *Typical cross-section through a one-way spanning solid slab*

3.10.2   Reinforcement areas

The requirements for minimum and maximum areas of main reinforcement are the same as those for beams given in Section 3.9.4.

It should be appreciated that since one-way spanning slabs are designed as a series of 1 m wide beams, the area of steel calculated is that required per metre width. The areas of round bar reinforcement spaced at various centres per metre width are given in Table 3.13.

Whilst for the purpose of design a slab may be considered as a series of 1 m wide beams, these will in fact be cast monolithically. Therefore additional reinforcement is included on top of and at right angles to the

Table 3.13   Areas of round bar reinforcement spaced at various centres (mm² per 1 m width)

Diameter (mm)	Spacing (mm)									
	75	100	125	150	175	200	225	250	275	300
6	377	283	226	188	162	141	126	113	103	94
8	670	503	402	335	287	251	223	201	183	168
10	1047	785	628	524	449	393	349	314	286	262
12	1508	1131	905	754	646	565	503	452	411	377
16	2681	2011	1 608	1340	1149	1005	894	804	731	670
20	4189	3142	2 513	2094	1795	1571	1396	1257	1142	1047
25	6546	4909	3 927	3272	2805	2454	2182	1963	1785	1636
32	—	8042	6 434	5362	4596	4021	3574	3217	2925	2681
40	—	—	10 053	8378	7181	6283	5585	5027	4570	4189

main reinforcement, as shown in Figure 3.26. This is called distribution steel and its purpose is to ensure distribution of the loading on to the main reinforcement.

The minimum area of distribution steel that must be provided is the same as for the main reinforcement. Normally the size of bars used in a slab should not be less than 10 mm diameter or more than 20 mm diameter.

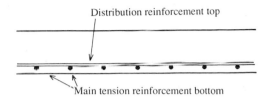

Distribution reinforcement top

Main tension reinforcement bottom

Figure 3.26 *Cross-section through a one-way spanning slab showing the position of the reinforcement*

3.10.3 Minimum spacing of reinforcement

The requirements for minimum spacing of reinforcement are the same as those for beams given in Section 3.9.5. However, for practical reasons the spacing of bars in a slab should usually be not less than 150 mm.

3.10.4 Maximum spacing of reinforcement

The clear distance between bars in a slab should never exceed the lesser of three times the effective depth ($3d$) or 750 mm.

Furthermore, unless crack widths are to be checked by direct calculation, the following rules must be complied with to ensure that crack widths do not exceed the maximum acceptable limit of 0.3 mm:

(a) No further check is required on bar spacing if:
  (i) Either grade 250 steel is used and $h \ngtr 250$ mm
  (ii) Or grade 460 steel is used and $h \ngtr 200$ mm
  (iii) Or the percentage reinforcement provided ($100A_s/bd$) is less than 0.3 per cent.
(b) If none of the conditions (i)–(iii) applies, then the bar spacing given in Table 3.14 should be used where the percentage of reinforcement contained in the slab is greater than 1 per cent. Table 3.14 is based on BS 8110 Part 1 Table 3.30.

Table 3.14 Clear distance between bars in slabs when percentage steel content is greater than 1 per cent

Steel characteristic strength f_y	250 N/mm²	460 N/mm²
Maximum clear distance between bars	300	160

(c) Where the percentage of steel contained in the slab is greater than 0.3 per cent but less than 1 per cent the spacing values given in Table 3.14 may be divided by the actual percentage figure. Thus if a slab contained 0.5 per cent steel with a yield stress $f_y = 250 \, \text{N/mm}^2$ then the maximum clear distance between bars, permitted by the code, would be $300/0.5 = 600 \, \text{mm}$. However, for practical reasons, the spacing of bars in a slab should not usually be more than 300 mm.

3.10.5 Bending ULS

Since a slab may be considered for design purposes to be a series of 1 m wide beams, its design ultimate resistance moment may be obtained by the same methods described for beams in Section 3.9.7, taking the breadth b as $1 \, \text{m} = 1000 \, \text{mm}$.

3.10.6 Cracking SLS

The rules relating to minimum bar areas and maximum spacing given in Sections 3.10.2 and 3.10.4 will ensure that crack widths do not exceed the general limit of 0.3 mm. However, when it is necessary to calculate specific crack width values, reference should be made to the guidance given in BS 8110 Part 2.

3.10.7 Deflection SLS

The deflection requirements for slabs are the same as those for beams given in Section 3.9.9.

3.10.8 Shear ULS

For practical reasons BS 8110 does not recommend the inclusion of shear reinforcement in solid slabs less than 200 mm deep. Therefore if no shear reinforcement is to be provided, the design shear stress v should not exceed the design ultimate shear stress v_c given in Table 3.12 of this manual. Thus for solid slabs up to 200 mm thick,

$$v = \frac{V}{bd} \ngtr v_c$$

This requirement will cater for the majority of one-way spanning slabs. However, for slabs thicker than 200 mm, where v is greater than v_c, shear reinforcement in the form recommended in BS 8110 Table 3.17 should be provided.

Let us now look at some examples on the design of simply supported one-way spanning slabs.

Example 3.10

A reinforced concrete slab is subjected to an ultimate moment M_u of 45 kN m per metre width, inclusive of its self-weight. The overall thickness of the slab is to be 200 mm, and 16 mm diameter main reinforcing bars are to be used. Using the simplified stress block formulae given in BS 8110 Part 1, check the adequacy of the slab thickness and determine the spacing for the main bars together with the size and spacing of the distribution bars for the following conditions:

Grade 40 concrete and mild steel reinforcement

Grade 35 concrete and high yield reinforcement

A 1 m width of slab will be considered for design purposes, as shown in Figure 3.27.

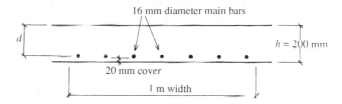

Figure 3.27   *Cross-section through slab considered for design*

First,

Effective depth d = overall depth h − (bar diameter/2) − cover = 200 − 8 − 20 = 172 mm

Grade 40 concrete, MS reinforcement
Grade 40 concrete $f_{cu} = 40$ N/mm^2
MS reinforcement $f_y = 240$ N/mm^2
Ultimate bending moment $M_u = 45$ kN m = 45 × 10^6 N mm

Use the BS 8110 simplified stress block formulae. First,

$$K = \frac{M}{bd^2 f_{cu}} = \frac{45 \times 10^6}{1000 \times 172^2 \times 40} = 0.038 < K' = 0.156$$

Therefore compression reinforcement is not necessary and the slab thickness is adequate.

The area of main tensile reinforcement required can now be calculated. The lever arm is given by

$$z = d[0.5 + \sqrt{(0.25 - K/0.9)}] = d[0.5 + \sqrt{(0.25 - 0.038/0.9)}] = 0.956d$$

But the lever arm depth must not be taken as greater than $0.95d$; therefore this limiting value will be used to calculate the area of tensile reinforcement required. Thus

$$A_s \text{ required} = \frac{M}{0.87 f_y z} = \frac{45 \times 10^6}{0.87 \times 250 \times 0.95 \times 172} = 1266 \text{ mm}^2 \text{ per metre width}$$

This area can be compared with the reinforcement areas given in Table 3.13 to enable suitable centres to be chosen for the 16 mm diameter bars specified:

Provide 16 mm diameter MS main bars at 150 mm centres (A_s per metre = 1340 mm^2).

The area of distribution reinforcement may now be determined. For mild steel reinforcement the minimum area to be provided is 0.24 per cent of the gross cross-sectional area of the slab. Therefore

$$\text{Minimum area required} = \frac{0.24}{100} \times 1000 \times 200 = 480 \text{ mm}^2 \text{ per metre run}$$

Hence by reference to Table 3.13:

Provide 10 mm diameter MS distribution bars at 150 mm centres (A_s per metre = 524 mm^2)

Check the maximum bar spacing needed to satisfy the cracking SLS. The overall depth $h = 200$ mm $\not> 250$ mm for MS reinforcement; therefore the clear distance between bars should not exceed the lesser of $3d$ or 750 mm. Hence the maximum clear distance between bars is $3 \times 172 = 516$ mm. Both the main and distribution bar spacing provided is therefore satisfactory.

Grade 35 concrete, HY reinforcement

Grade 35 concrete $f_{cu} = 35$ N/mm^2

HY reinforcement $f_y = 460$ N/mm^2

$$K = \frac{45 \times 10^6}{1000 \times 172^2 \times 35} = 0.043 < K' = 0.156$$

$$z = d[0.5 + \sqrt{(0.25 - 0.043/0.9)}] = 0.95d \not> 0.95d$$

$$A_s \text{ required} = \frac{45 \times 10^6}{0.87 \times 460 \times 0.95 \times 172} = 688 \text{ mm}^2 \text{ per metre width}$$

Provide 16 mm diameter HY main bars at 275 mm centres (A_s per metre = 731 mm^2).

The minimum area of HY distribution reinforcement to be provided is 0.13 per cent of the gross cross-sectional area of the slab. Therefore

$$\text{Minimum area required} = \frac{0.13}{100} \times 1000 \times 200 = 260 \text{ mm}^2 \text{ per metre run}$$

Provide 10 mm diameter HY distribution bars at 300 centres (A_s per metre = 262 mm^2).

Check the maximum bar spacing needed to satisfy the cracking SLS. The overall depth $h = 200$ mm $\not> 200$ mm for HY reinforcement; therefore the clear distance between bars should again not exceed the lesser of $3d$ or 750 mm. Hence the maximum clear distance between bars will again be 516 mm, and therefore the spacing of both the main and distribution bars is satisfactory.

Example 3.11

A 250 mm thick simply supported reinforced concrete slab spans 5 m. Design a suitable slab using grade 40 concrete and high yield reinforcement to support the following characteristic loads:

Imposed 4.0 kN/m^2
Finishes 0.5 kN/m^2
Concrete 24 kN/m^3

The slab will be in a mild exposure situation.

Consider a 1 m width of slab as shown in Figure 3.28. In this example it is necessary first to calculate the ultimate design load and ultimate bending moment.

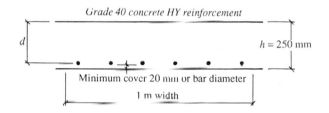

Figure 3.28 *Cross-section through slab considered for design*

Loading
Characteristic imposed load = 4 kN/m^2
Characteristic imposed UDL $Q_k = 4 \times 5 \times 1 = 20$ kN

Characteristic dead load: finishes   0.5
                   self-weight 24×0.25 6.0
                   total   6.5 kN/m^2

Characteristic dead UDL $G_k = 6.5 \times 5 \times 1 = 32.5$ kN

ULS imposed UDL $\gamma_f Q_k = 1.6 \times 20$ 32.0 kN
ULS dead   UDL $\gamma_f G_k = 1.4 \times 32.5$ 45.5 kN
ULS total   UDL   77.5 kN

The slab load diagram is shown in Figure 3.29.

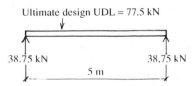

Ultimate design UDL = 77.5 kN

38.75 kN   38.75 kN

5 m

Figure 3.29 *Slab load diagram*

Bending

$$M_u = \frac{WL}{8} = \frac{77.5 \times 5}{8} = 48.44 \, kN\,m = 48.44 \times 10^6 \, N\,mm$$

Use the BS 8110 simplified stress block formulae. First,

$$K = \frac{M}{bd^2 f_{cu}} \leqslant K' = 0.156$$

Let us assume that 12 mm diameter main bars will be used with a cover of 20 mm. Hence the effective depth provided is

$$d = 250 - 20 - \frac{12}{2} = 224 \, mm$$

Therefore

$$K = \frac{48.44 \times 10^6}{1000 \times 224^2 \times 40} = 0.024 < K' = 0.156$$

$$z = d[0.5 + \sqrt{(0.25 - K/0.9)}]$$
$$= d[0.5 + \sqrt{(0.25 - 0.024/0.9)}] = 0.97d > 0.95d$$

Therefore use 0.95d. Next,

$$A_s \text{ required} = \frac{M}{0.87 f_y z} = \frac{48.44 \times 10^6}{0.87 \times 460 \times 0.95 \times 224} = 569 \, mm^2 \text{ per metre width}$$

Provide 12 mm diameter HY main bars at 175 mm centres (A_s per metre = 646 mm^2).

Minimum area of distribution steel $= \dfrac{0.13}{100} \times 1000 \times 250 = 325 \, mm^2$ per metre run.

Provide 10 mm diameter HY distribution bars at 225 mm centres (A_s per metre = 349 mm^2).

Cracking

Check the maximum bar spacing needed to satisfy the cracking SLS. The overall depth $h = 250 > 200$ mm for HY reinforcement. However, the percentage of main reinforcement provided is not greater than 0.3 per cent:

$$\frac{100 A_s}{bd} = \frac{100 \times 646}{1000 \times 224} = 0.288 < 0.3 \text{ per cent}$$

Therefore the clear distance between bars should not exceed the lesser of $3d = 3 \times 224 = 672$ mm or 750 mm. Therefore both the main and distribution bar spacing provided is satisfactory.

Shear

Check the shear ULS. The ultimate design shear force (at support) is

$$V = \frac{\text{ultimate UDL}}{2} = \frac{77.5}{2} = 38.75 \, kN$$

The maximum design shear stress occurring is

$$v = \frac{V}{b_v d} = \frac{38.75 \times 10^3}{1000 \times 224} = 0.173 \text{ N/mm}^2$$

Now

$$\frac{100 A_s}{b_v d} = 0.288 \text{ per cent}$$

Thus the design concrete shear stress (from Table 3.12) v_c is 0.48 N/mm^2 for grade 25 concrete. Therefore for grade 40 concrete,

$$v_c = 0.48 (f_{cu}/25)^{1/3} = 0.48 (40/25)^{1/3} = 0.56 \text{ N/mm}^2$$

Hence since $v = 0.173$ N/mm^2 is less than $v_c = 0.56$ N/mm^2, no shear reinforcement is required.

Deflection

Check the deflection SLS by reference to the recommended span to depth ratios given in Table 3.9 of this manual. The basic span to effective depth ratio is 20. But this must be modified by the factor for the amount of tension reinforcement in the slab, obtained from Table 3.10, when

$$\frac{M}{bd^2} = \frac{48.44 \times 10^6}{1000 \times 224^2} = 0.97$$

From the expression given in the second footnote to Table 3.10,

$$f_s = \frac{5}{8} f_y \frac{A_{s\,req}}{A_{s,\,prov}} = \frac{5}{8} \times 460 \times \frac{569}{646} = 253.23 \text{ N/mm}^2$$

Therefore the modification factor for tension reinforcement is 1.56. The allowable span to effective depth ratio is $20 \times 1.56 = 31.2$, and the actual span to effective depth ratio is $5000/224 = 22.32 < 31.2$. Hence the slab is adequate in deflection.

3.11   Columns

Reinforced concrete columns are classified in BS 8110 as either unbraced or braced. The difference relates to the manner by which lateral stability is provided to the structure as a whole. A concrete framed building may be designed to resist lateral loading, such as that resulting from wind action, in two distinct ways:

(a) The beam and column members may be designed to act together as a rigid frame in transmitting the lateral forces down to the foundations (Figure 3.30). In such an instance the columns are said to be unbraced and must be designed to carry both the vertical (compressive) and lateral (bending) loads.

(b) Alternatively the lateral loading may be transferred via the roof and floors to a system of bracing or shear walls, designed to transmit the resulting forces down to the foundations (Figure 3.31). The columns are then said to be braced and consequently carry only vertical loads.

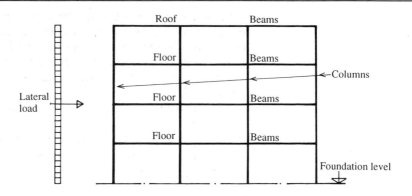

Figure 3.30 *Unbraced frame; lateral load must be resisted and transmitted down to the foundations by interaction between the beam and column frame members*

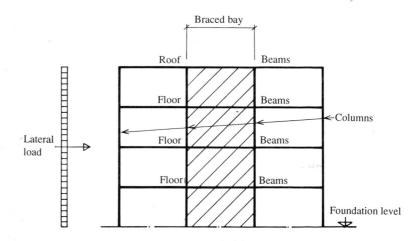

Figure 3.31 *Braced frame; lateral load transmitted down to foundations through a system of bracing or shear walls*

Columns are further classified in BS 8110 as either short or slender. A braced column may be considered to be short when neither of its effective height ratios exceeds 15; that is, for a short braced column

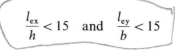

$$\frac{l_{ex}}{h} < 15 \quad \text{and} \quad \frac{l_{ey}}{b} < 15$$

where

l_{ex} effective height in respect of column major axis
l_{ey} effective height in respect of column minor axis
  h depth in respect of major axis
  b width in respect of minor axis

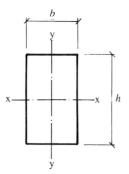

Figure 3.32 *Cross-section through a rectangular column*

The lateral dimensions h and b relative to the axes of a rectangular column are shown in Figure 3.32.

If either of the effective height ratios did exceed 15 then the braced column would be considered to be a slender column. In such circumstances the slender braced column would have to be designed to resist additional bending moments induced by lateral deflection.

For the purpose of this manual only the design of short braced columns will be studied.

The effective heights l_{ex} and l_{ey} about the respective axes are influenced by the degree of fixity at each end of the column. Simplified recommendations are given in BS 8110 Part 1 for the assessment of effective column heights for common situations. For braced columns the effective height is obtained by multiplying the clear height between restraints l_o by an end condition factor β from BS 8110 Part 1 Table 3.21, reproduced here as Table 3.15:

$$\text{Effective heights } l_{ex} \text{ or } l_{ey} = \beta l_o$$

Table 3.15 Values of β for braced columns (BS 8110 Part 1 1985 Table 3.21)

End condition at top	End condition at bottom		
	1	2	3
1	0.75	0.80	0.90
2	0.80	0.85	0.95
3	0.90	0.95	1.00

The types of end condition that influence end fixity are defined in BS 8110 as follows:

Condition 1 The end of the column is connected monolithically to beams on either side which are at least as deep as the overall dimension of the column in the plane considered (Figure 3.33). Where the column is connected to a foundation structure, this should be of a form specifically designed to carry moment.

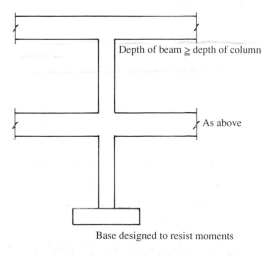

Figure 3.33 *End fixity condition 1*

Condition 2 The end of the column is connected monolithically to beams or slabs on either side which are shallower than the overall dimensions of the column in the plane considered (Figure 3.34).

Condition 3 The end of the column is connected to members which, while not specifically designed to provide restraint to rotation of the column will nevertheless provide some nominal restraint (Figure 3.35).

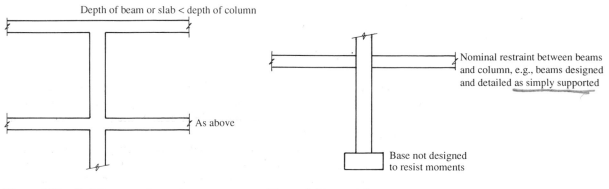

Depth of beam or slab < depth of column

As above

Nominal restraint between beams and column, e.g., beams designed and detailed as simply supported

Base not designed to resist moments

Figure 3.34 *End fixity condition 2*   **Figure 3.35** *End fixity condition 3*

Where a more accurate assessment of the effective height is desired it may be calculated from the equations given in Section 2.5 of BS 8110 Part 2. The basic mode of failure of a braced short column is by crushing of the constituent materials due to the compressive loads.

The various aspects of the design of braced short columns, including a number of dimensional considerations which can influence the design, will be considered under the following headings:

(a) Column cross-section
(b) Main reinforcement areas
(c) Minimum spacing of reinforcement
(d) Maximum spacing of reinforcement
(e) Lateral reinforcement
(f) Compressive ULS
(g) Shear ULS
(h) Cracking SLS
(i) Lateral deflection.

3.11.1 Column cross-section

The provisions of column design given in BS 8110 apply to vertical load bearing members whose greater cross-sectional dimension does not exceed four times its smaller dimension. This proviso is illustrated in Figure 3.36. It should be appreciated that square, circular or any other symmetrical shape will satisfy this requirement.

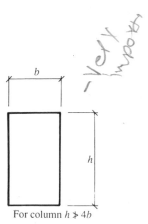

For column $h \not> 4b$

Figure 3.36 *Cross-sectional limitation for columns*

A vertical load bearing member whose breadth exceeds four times its thickness is classified as a wall and should be designed in accordance with the provisions for reinforced concrete walls.

Initially the cross-sectional dimensions may be determined by taking into account the durability, fire resistance and slenderness requirements. It is suggested for practical reasons appertaining to the *in situ* casting of columns that the minimum lateral dimension should not be less than 200 mm.

3.11.2 Main reinforcement areas

Sufficient reinforcement must be provided in order to control cracking of the concrete. Therefore the minimum area of compression reinforcement in a column should not be less than 0.4 per cent of the total concrete area, irrespective of the type of steel.

A maximum steel content is also specified to ensure proper placing and compaction of concrete around reinforcement. Therefore the maximum area of compression reinforcement in a vertically cast column should not exceed 6 per cent of the gross cross-sectional area. If it is necessary to lap the compression bars in a column, as shown in Figure 3.37, the maximum area limit may be increased to 10 per cent at lap positions.

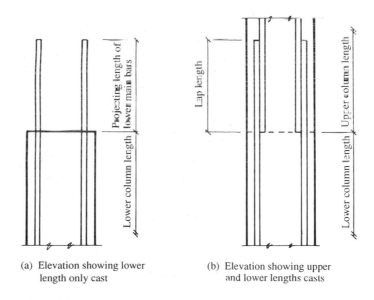

(a) Elevation showing lower length only cast

(b) Elevation showing upper and lower lengths casts

Figure 3.37 *Lapped compression bars in a column*

For practical reasons the minimum number of longitudinal bars should be four in a square or rectangular column and six in a circular column. Their minimum size should be 12 mm diameter. The areas of round bar reinforcement have already been given in Table 3.8 in connection with the design of beams.

The general requirements relating to the main reinforcement in columns are illustrated and summarized in Figure 3.38, to which the following symbols apply:

A_g gross cross-sectional area of the column

A_{sc} area of main longitudinal reinforcement

A_c net cross-sectional area of concrete: $A_c = A_g - A_{sc}$

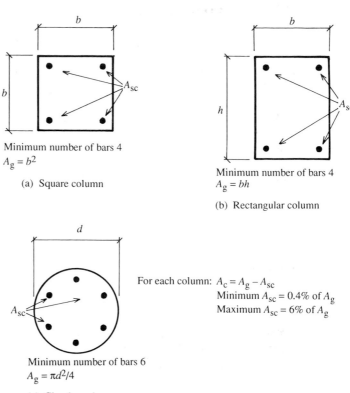

(a) Square column

Minimum number of bars 4
$A_g = b^2$

(b) Rectangular column

Minimum number of bars 4
$A_g = bh$

For each column: $A_c = A_g - A_{sc}$
Minimum $A_{sc} = 0.4\%$ of A_g
Maximum $A_{sc} = 6\%$ of A_g

(c) Circular column

Minimum number of bars 6
$A_g = \pi d^2/4$

Figure 3.38 *Reinforcement requirements for columns*

In this context it is possible that confusion could arise with respect to the symbol A_c when reading BS 8110. It is defined in clause 3.8.1.1 as the net cross-sectional area of concrete in a column, whereas in clause 3.12.5.2, relating to minimum reinforcement areas, A_c is defined as the total area of concrete. Therefore to avoid confusion here A_c has been taken to be the net cross-sectional area of concrete, and the symbol A_g has been adopted for the gross or total cross-sectional area of the column.

3.11.3   Minimum spacing of reinforcement

The minimum spacing of main reinforcement in a column is the same as that given for beams in Section 3.9.5.

3.11.4 Maximum spacing of reinforcement

No maximum spacing is stipulated in BS 8110 for the main bars in a column other than those implied by the recommendations for containment of compression reinforcement. However, for practical reasons it is considered that the maximum spacing of main bars should not exceed 250 mm.

3.11.5 Lateral reinforcement

Lateral reinforcement in columns is commonly referred to as links or ties or sometimes binders. Its purpose is to prevent lateral buckling of the longitudinal main bars due to the action of compressive loading, and the subsequent spalling of the concrete cover. This is illustrated in Figure 3.39.

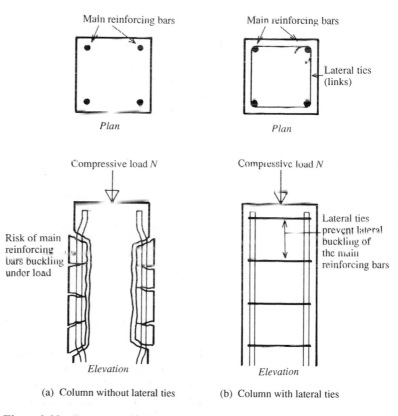

(a) Column without lateral ties

(b) Column with lateral ties

Figure 3.39 *Function of lateral ties*

The diameter of lateral ties must not be less than one-quarter the size of the largest main compression bar and in no case less than 6 mm. The maximum spacing of lateral ties must not be more than twelve times the diameter of the smallest main compression bar. Furthermore it is common practice to ensure that the spacing never exceeds the smallest cross-sectional dimension of the column.

A main compression bar contained by a link passing around it and having an internal angle of not more than 135° is said to be restrained. BS 8110 stipulates that every corner bar and each alternate bar should be restrained. Intermediate bars may be unrestrained provided that they are not more than 150 mm away from a restrained bar.

The compression bars in a circular column will be adequately restrained if circular shaped links are provided passing around the main bars.

Lateral reinforcement arrangements to satisfy these requirements are illustrated in Figure 3.40.

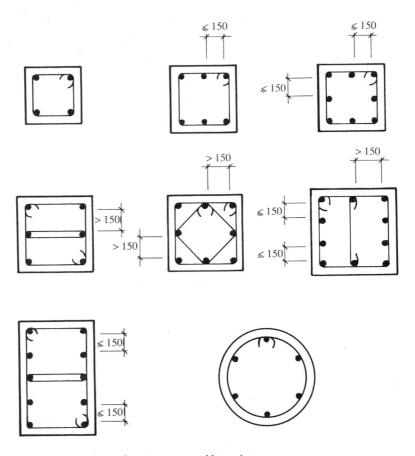

Figure 3.40 *Typical arrangement of lateral ties*

3.11.6   Compressive ULS

The compressive ULS analysis for short braced columns given in BS 8110 may basically be divided into three categories:

(a) Short braced axially loaded columns.
(b) Short braced columns supporting an approximately symmetrical arrangement of beams.

(c) Short braced columns supporting vertical loads and subjected to either uniaxial or biaxial bending.

Each of these categories will be discussed in turn.

Short braced axially loaded columns

When a short braced column supports a concentric compressive load or where the eccentricity of the compressive load is nominal, it may be considered to be axially loaded. Nominal eccentricity in this context is defined as being not greater than 0.05 times the overall column dimension in the plane of bending or 20 mm. Thus for a lateral column dimension not greater than 400 mm the value 0.05 times the dimension would apply, and over 400 mm the 20 mm limit would apply. These load conditions are illustrated in Figure 3.41.

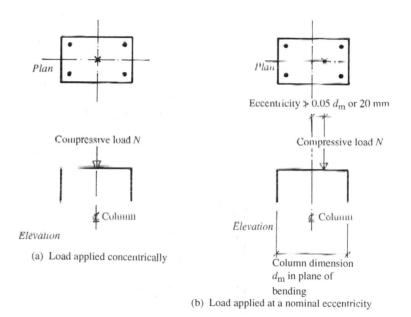

Plan

Eccentricity $\gg 0.05\ d_m$ or 20 mm

Compressive load N

$\cancel{\text{C}}$ Column

Elevation

(a) Load applied concentrically

Plan

Compressive load N

$\cancel{\text{C}}$ Column

Elevation

Column dimension d_m in plane of bending

(b) Load applied at a nominal eccentricity

Figure 3.41 *Axially loaded columns*

The ultimate compressive load for columns in such instances is obtained from the following expression, which includes an allowance for the material partial safety factory γ_m:

BS 8110 equation 38: $\qquad N = 0.4 f_{cu} A_c + 0.80 A_{sc} f_y$

where

A_c net cross-sectional area of concrete in a column (excluding area of reinforcement)

A_{sc} area of vertical reinforcement

f_{cu} characteristic strength of concrete

f_y characteristic strength of reinforcement

N design ultimate axial load on column

Now $A_c = A_g - A_{sc}$, where A_g is the gross cross-sectional area of the column. Hence by substituting this value in the BS 8110 expression 38 it becomes:

Equation 38(a): $N = 0.4f_{cu}(A_g - A_{sc}) + 0.75A_{sc}f_y$

It should be appreciated that the two parts of this expression represent the load sustained by each of the column's two composite materials, concrete and steel. That is,

> Ultimate load supported by the concrete
> = ultimate concrete design stress × net concrete area
> $= 0.4f_{cu}A_c = 0.4f_{cu}(A_g - A_{sc})$
> Ultimate load supported by the steel
> = ultimate steel design stress × steel area
> $= 0.75f_y A_{sc}$

Short braced columns supporting an approximately symmetrical arrangement of beams

For columns within this category, either the columns must support symmetrical beam arrangements (Figure 3.42), or the span of the beams on adjacent sides of the column must not differ by more than 15 per cent of the longer span (Figure 3.43). Furthermore the column must only support beams carrying uniformly distributed loads.

Provided that these conditions are met, a column may be designed as axially loaded using the following modified expression, which again includes an allowance for the material partial safety factor γ_m:

BS 8110 equation 39: $N = 0.35f_{cu}A_{sc} + 0.67A_{sc}f_y$

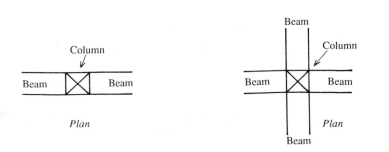

Figure 3.42   *Columns supporting symmetrical beam arrangements*

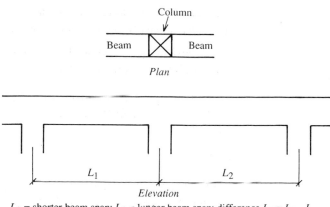

Plan

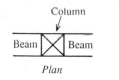

Elevation

L_1 = shorter beam span; L_2 = longer beam span; difference $L_d = L_2 - L_1$
Percentage difference in spans = L_d/L_1 × 100 ≯ 15%

Figure 3.43 *Column supporting beams of differing spans where the difference is not greater than 15 per cent*

The terms have the same definition as those in the previous category, and again by substituting $A_c = A_g - A_{sc}$ in the expression it becomes:

Equation 39(a): $N = 0.35f_{cu}(A_g - A_{sc}) + 0.67A_{sc}f_y$

Short braced columns supporting vertical loads and subjected to either uniaxial or biaxial bending

In addition to vertical loading, columns supporting beams on adjacent sides whose spans vary by more than 15 per cent will be subjected to uniaxial bending, that is bending about one axis. Such an arrangement is shown in Figure 3.44.

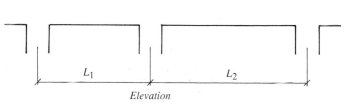

Plan

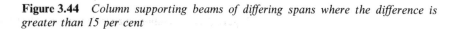

Elevation

L_1 = shorter beam span; L_2 = longer beam span; difference $L_d = L_2 - L_1$
Percentage difference in spans = L_d/L_1 × 100 > 15%

Figure 3.44 *Column supporting beams of differing spans where the difference is greater than 15 per cent*

Furthermore, columns around the outside of a building are often, owing to the configuration of the beams they support, subjected to biaxial bending as shown in Figure 3.45. In such instances the columns should be designed to resist bending about both axes. However, when such columns are symmetrically reinforced, BS 8110 Part 1 allows the adoption of a simplified analysis. The approach is to design the columns for an increased moment about one axis only, using the following procedure in relation to Figure 3.46. When

$$\frac{M_x}{h'} \geqslant \frac{M_y}{b'}$$

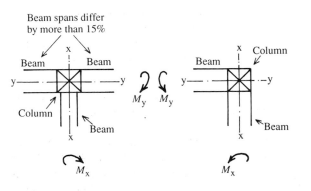

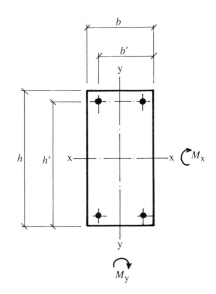

Figure 3.45 *Columns supporting beam arrangements that produce biaxial bending*

Figure 3.46 *Plan on column subject to biaxial bending*

the increased moment about the x–x axis is:

BS 8110 equation 40: $M'_x = M_x + \dfrac{\beta h'}{b'} M_y$

When

$$\frac{M_x}{h'} < \frac{M_y}{b'}$$

the increased moment about the y–y axis is:

BS 8110 equation 41: $M'_y = M_y + \dfrac{\beta b'}{h'} M_x$

where

b overall section dimension perpendicular to $y-y$ axis

b' effective depth perpendicular to $y-y$ axis

h overall section dimension perpendicular to $x-x$ axis

h' effective depth perpendicular to $x-x$ axis

M_x bending moment about $x-x$ axis

M_y bending moment about $y-y$ axis

β coefficient obtained from BS 8110 Part 1 Table 3.24, reproduced here as Table 3.16

Table 3.16 Values of the coefficient β (BS 8110 Part 1 1985 Table 3.24)

N/bhf_{cu}	0	0.1	0.2	0.3	0.4	0.5	$\geqslant 0.6$
β	1.00	0.88	0.77	0.65	0.53	0.42	0.30

Having established the increased moment about one of the column axes, the section can then be designed for the combination of vertical load and bending.

Design charts for the design of symmetrically reinforced columns subject to vertical loads and bending are presented in BS 8110 Part 3. There is a separate chart for each grade of concrete combined with HY reinforcement and individual d/h ratios. The area of reinforcement can be found from the appropriate chart using the N/bh and M/bh^2 ratios for the column section being designed. Chart 38 is reproduced here as Figure 3.47.

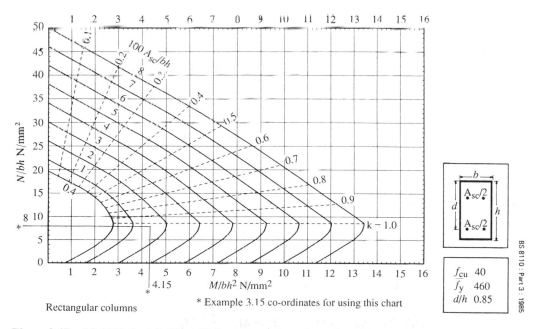

Rectangular columns

* Example 3.15 co-ordinates for using this chart

BS 8110 : Part 3 : 1985

f_{cu}	40
f_y	460
d/h	0.85

Figure 3.47 *BS 8110 Part 3 Chart 38 for rectangular column design*

3.11.7   Shear ULS

Axially loaded columns are not subjected to shear and therefore no check is necessary.

Rectangular columns subjected to vertical loading and bending do not need to be checked for shear when the ratio of the moment to the vertical load M/N is less than $0.75h$. However, this is only provided that the shear stress does not exceed the lesser of $0.8\sqrt{f_{cu}}$ or $5\,\text{N/mm}^2$; if it does, the size of the column section would have to be increased.

3.11.8   Cracking SLS

Since cracks are produced by flexure of the concrete, short columns that support axial loads alone do not require checking for cracking. Furthermore, it is advised in BS 8110 that cracks due to bending are unlikely to occur in columns designed to support ultimate axial loads greater than $0.2f_{cu}A_c$. All other columns subject to bending should be considered as beams for the purpose of examining the cracking SLS.

3.11.9   Lateral deflection

No deflection check is necessary for short braced columns. When for other types of column the deflection needs to be checked, reference should be made to Section 3 of BS 8110 Part 2 for guidance.

3.11.10   Design summary for concrete columns

The design procedure for short braced concrete columns may be summarized as follows:

(a) Ensure that the column satisfies the requirements for braced columns.
(b) Ensure that the column is a short column by reference to its slenderness ratio.
(c) (i) If the column is axially loaded, design for the compressive ULS using BS 8110 equation 38.
   (ii) If the column supports an approximately symmetrical arrangement of beams, design for the compressive ULS using BS 8110 equation 39.
   (iii) If the column is subjected to either uniaxial or biaxial bending, design for the combined ULS of compression and bending by reference to BS 8110 Part 3 design charts.
(d) Check shear ULS for columns subjected to vertical loading and bending. No check is necessary for axially loaded columns.
(e) Check cracking SLS for columns subjected to vertical loading and bending. No check is necessary for axially loaded columns.

Let us now look at some examples on the design of short braced columns.

Example 3.12 ~~478.02kN~~   ~~300 × 300~~

A short braced column in a situation of mild exposure supports an ultimate axial load of 1000 kN, the size of the column being 250 mm × 250 mm. Using grade 30 ~~15~~ concrete with mild steel reinforcement, calculate the size of all reinforcement required and the maximum effective height for the column if it is to be considered as a short column.

Since the column is axially loaded, equation 38(a) will apply:

$$N = 0.4f_{cu}(A_g - A_{sc}) + 0.75A_{sc}f_y$$
$$1000 \times 10^3 = 0.4 \times 30[(250 \times 250) - A_{sc}] + 0.75A_{sc} \times 250$$
$$1\,000\,000 - 750\,000 - 12A_{sc} + 187.5A_{sc}$$

Hence

$$A_{sc} \text{ required} = \frac{250\,000}{175.5} = 1424.5\,mm^2$$

This area can be compared with the reinforcement areas given in Table 3.8 to enable suitable bars to be selected:

Provide four 25 mm diameter MS bars ($A_{sc} = 1966\,mm^2$).

Now determine the size and pitch needed for the lateral ties. The diameter required is the greater of (a) one-quarter of the diameter of the largest main bar, that is $25/4 = 6.25$ mm, or (b) 6 mm. The pitch required is the lesser of (a) 12 times the diameter of the smallest main bar, that is $12 \times 25 = 300$ mm, or (b) the smallest cross-sectional dimension of column, that is 250 mm. Thus:

Provide 8 mm diameter MS links at 250 mm pitch.

Now the maximum effective height ratio l_e/h for a short braced column is 15. Hence the maximum effective height is

$$l_e = 15h = 15 \times 250 = 3750\,mm$$

Thus the maximum effective height for this column to be considered as a short column would be 3.75 m.

A cross-section through the finished column is shown in Figure 3.48.

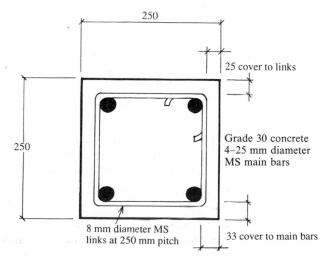

Figure 3.48 *Cross-section through finished column*

Example 3.13

A short braced reinforced concrete column is required to support an ultimate axial load of 1300 kN. Assuming 2 per cent main steel, calculate the diameter of circular column required and choose suitable MS main bars if grade 30 concrete is used.

Since the column is axially loaded, equation 38(a) will apply:

$$N = 0.4 f_{cu}(A_g - A_{sc}) + 0.75 A_{sc} f_y$$

In this example both the size of column and the area of reinforcement are unknown. However, a 2 per cent steel content may be assumed; therefore A_{sc} is 2 per cent of A_g or $0.02 A_g$. By substituting this in the expression, only one unknown A_g remains:

$$1300 \times 10^3 = 0.4 \times 30(A_g - 0.02 A_g) + 0.75 \times 0.02 A_g \times 250$$
$$1300 \times 10^3 = 12 A_g - 0.24 A_g + 3.75 A_g = 15.51 A_g$$
$$A_g = \frac{1300 \times 10^3}{15.51} = 83\,816.89\,\text{mm}^2$$

Since the column is circular,

$$A = \frac{\pi d^2}{4} = 83\,816.89$$

$$d = \sqrt{\left(\frac{4 \times 83\,816.89}{\pi}\right)} = 326.7\,\text{mm}$$

Provide a 330 mm diameter grade 30 concrete circular column.

The actual A_g is $\pi 330^2/4 = 85\,529.86\,\text{mm}^2$. Therefore if 2 per cent steel content is to be provided,

$$\text{Area of main bars} = 2 \text{ per cent of } A_g = \frac{85\,529.86}{50} = 1711\,\text{mm}^2$$

Provide six 20 mm diameter MS bars ($A_{sc} = 1884\,\text{mm}^2$).

Example 3.14

A short braced reinforced concrete column supports an approximately symmetrical arrangement of beams which result in a total ultimate vertical load of 1500 kN being applied to the column. Assuming the percentage steel content to be 1 per cent, choose suitable dimensions for the column and the diameter of the main bars. Use grade 35 concrete with HY reinforcement in a square column.

Since the column supports an approximately symmetrical arrangement of beams, we will assume that their spans do not differ by 15 per cent and hence equation 39(a) will apply:

$$N = 0.35 f_{cu}(A_g - A_{sc}) + 0.67 A_{sc} f_y$$

Now A_{sc} is 1 per cent of A_g or $0.01A_g$. Therefore

$$1500 \times 10^3 = 0.35 \times 35(A_g - 0.01A_g) + 0.67 \times 0.01A_g \times 460$$
$$1500 \times 10^3 = 12.25A_g - 0.1225A_g + 3.082A_g = 15.21A_g$$
$$A_g = \frac{1500 \times 10^3}{15.21} = 98\,619.33\,\text{mm}^2$$

Since the column is square, the length of side is $\sqrt{98\,619.33} = 314.04\,\text{mm}$.

Provide a 315 mm × 315 mm square grade 35 concrete column.

The actual A_g is $315 \times 315 = 99\,225\,\text{mm}^2$. Therefore

Area of main bars — 1 per cent of $A_g = \dfrac{99\,225}{100} = 992.25\,\text{mm}^2$

Provide four 20 mm diameter HY bars ($A_{sc} = 1256\,\text{mm}^2$).

Example 3.15

A short braced column supporting a vertical load and subjected to biaxial bending is shown in Figure 3.49. If the column is formed from grade 40 concrete, determine the size of HY main reinforcement required.

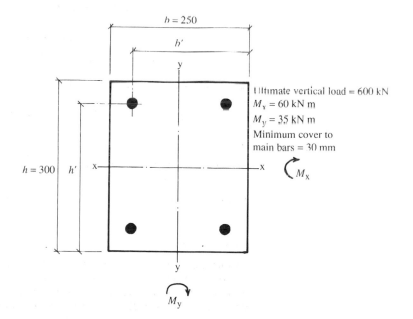

Figure 3.49 *Column subject to biaxial bending*

Since this column is subjected to bending it will be designed using the relevant BS 8110 Part 3 chart. To do so, it is first necessary to convert the biaxial bending into uniaxial bending by increasing one of the moments in accordance with the simplified procedure given in BS 8110 Part 1 as follows.

Assume 20 mm diameter bars will be adopted. Then $h' = 300 - 40 = 260$ and $b' = 250 - 40 = 210$. Thus

$$\frac{M_x}{h'} = \frac{60}{260} = 0.231 \quad \text{and} \quad \frac{M_y}{b'} = \frac{35}{210} = 0.17$$

Hence $M_x/h' > M_y/b'$. Therefore BS 8110 equation 40 will apply, where β is obtained from Table 3.16 using

$$\frac{N}{bhf_{cu}} = \frac{600 \times 10^3}{250 \times 300 \times 40} = 0.2$$

Hence from Table 3.16, $\beta = 0.77$. Therefore the increased moment about the x–x axis is given by

$$M_x' = 60 + 0.77 \times \frac{260}{210} \times 35 = 60 + 33.37 = 93.37 \, \text{kN m}$$

In order to determine which of the BS 8110 Part 3 charts to use we need to know the d/h ratio together with the f_{cu} and f_y values. f_{cu} is 40, f_y is 460 and $d/h = 260/300 = 0.87 \simeq 0.85$. Therefore we use BS 8110 Chart 38, reproduced earlier as Figure 3.47. To use the chart the following ratios must be calculated:

$$\frac{N}{bh} = \frac{600 \times 10^3}{250 \times 300} = 8 \quad \text{and} \quad \frac{M}{bh^2} = \frac{93.37 \times 10^6}{250 \times 300^2} = 4.15$$

From the chart, $100A_{sc}/bh = 1.6$. Therefore

$$A_{sc} = \frac{1.6bh}{100} = \frac{1.6 \times 250 \times 300}{100} = 1200 \, \text{mm}^2$$

Provide four 20 mm diameter HY bars ($A_{sc} = 1256 \, \text{mm}^2$).

3.12   References

BS 5328 1990 Concrete, Parts 1, 2, 3 and 4.

BS 8110 1985 Structural use of concrete.
  Part 1 Code of practice for design and construction.
  Part 2 Code of practice for special circumstances.
  Part 3 Design charts for singly reinforced beams, doubly reinforced beams and rectangular columns.

Manual for the Design of Reinforced Concrete Building Structures. Institution of Structural Engineers, October 1985.

Standard Method of Detailing Structural Concrete. Institution of Structural Engineers, August 1989.

For further information contact:

British Cement Asssociation (BCA), Wexham Springs, Slough, SL3 6PL.

4   Masonry elements

4.1   Structural design of masonry

The structural design of masonry is carried out in accordance with the guidance given in BS 5628 'Code of practice for use of masonry'. This is divided into the following three parts:

Part 1   Structural use of unreinforced masonry.
Part 2   Structural use of reinforced and prestressed masonry.
Part 3   Materials and components, design and workmanship.

The design of masonry dealt with in this manual is based on Part 1, which gives design recommendations for unreinforced masonry constructed of bricks, concrete blocks or natural stone.

When an unreinforced wall is found to be inadequate, consideration may be given to adding reinforcement or even prestressing the masonry. In such circumstances the calculations would be based upon the recommendations given in Part 2 of the code.

Guidance is given in the code on the design of walls to resist lateral loading, such as that resulting from wind loads, as well as vertical loading. However, this manual will concentrate on the design of vertically loaded walls.

4.2   Symbols

Those symbols used in BS 5628 that are relevant to this manual are as follows:

A horizontal cross-sectional area
b width of column
e_x eccentricity at top of a wall
f_k characteristic compressive strength of masonry
G_k characteristic dead load
g_A design vertical load per unit area
g_d design vertical dead load per unit area
h clear height of wall or column between lateral supports
h_{ef} effective height of wall or column
K stiffness coefficient
L length
l_{ef} effective length of wall
Q_k characteristic imposed load
t overall thickness of a wall or column
t_{ef} effective thickness of a wall or column
t_p thickness of a pier
t_1 thickness of leaf 1 of a cavity wall

t_2 thickness of leaf 2 of a cavity wall
β capacity reduction factor for walls and columns allowing for effects of slenderness and eccentricity
γ_f partial safety factor for load
γ_m partial safety factor for material

4.3 Definitions

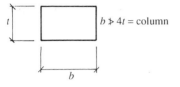

Figure 4.1 *Definition of a column*

The following definitions which are relevant to this manual have been abstracted from BS 5628 Part 1:

Column   An isolated vertical load bearing member whose width is not more than four times its thickness, as illustrated in Figure 4.1.

Effective height or length   The height or length of a wall, pier or column assumed for calculating the slenderness ratio.

Effective thickness   The thickness of a wall, pier or column assumed for calculating the slenderness ratio.

Lateral support   The support, in relation to a wall or pier, which will restrict movement in the direction of the thickness of the wall or, in relation to a column, which will restrict movement in the direction of its thickness or width. Lateral supports may be horizontal or vertical.

Load bearing walls   Walls primarily designed to carry an imposed vertical load in addition to their own weight.

Masonry   An assemblage of structural units, either laid *in situ* or constructed in prefabricated panels, in which the structural units are bonded and solidly put together with mortar or grout. Masonry may be reinforced or unreinforced.

Pier   A member which forms an integral part of a wall, in the form of a thickened section placed at intervals along the wall.

Slenderness ratio   The ratio of the effective height or effective length to the effective thickness.

Structural units   Bricks or blocks, or square dressed natural stone.

Single leaf wall   A wall of bricks or blocks laid to overlap in one or more directions and set solidly in mortar.

Double leaf (collar jointed) wall   Two parallel single leaf walls, with a space between not exceeding 25 mm, filled solidly with mortar and so tied together as to result in common action under load.

Cavity wall   Two parallel single leaf walls, usually at least 50 mm apart, and effectively tied together with wall ties, the space between being left as a continuous cavity or filled with non-load-bearing material.

Faced wall   A wall in which the facing and backing are so bonded as to result in common action under load.

Veneered wall   A wall having a facing which is attached to the backing, but not so bonded as to result in common action under load.

4.4 Materials

The fundamental properties of the individual materials that comprise a masonry wall are well understood and documented. Sadly, however, a designer's intentions may sometimes be frustrated by a lack of understanding of their combined behaviour. To use masonry successfully the designer must select bricks or blocks of appropriate quality, choose suitable mortar, specify their use correctly and devise appropriate details.

It is pointed out in Part 1 of the code that wall thicknesses derived from strength considerations may be insufficient to satisfy other performance requirements. Reference should therefore be made to BS 5628 Part 3 for guidance on such matters as durability, fire resistance, thermal insulation, sound insulation, resistance to damp penetration and provision for thermal movement, together with material, component and workmanship specification matters.

The main constituent materials and components used in the construction of masonry walls are as follows:

(a) Bricks

(b) Blocks

(c) Mortar

(d) Wall ties

(e) Damp proof courses.

Each will now be discussed in more detail.

4.4.1 Bricks

Bricks are walling units not exceeding 337.5 mm in length, 225 mm in width and 112.5 mm in height. They are produced from a range of materials, such as clay, concrete and sometimes a mixture of lime and sand or crushed stone. The mixture types are referred to as either calcium silicate bricks or sand lime bricks.

The standard format of clay bricks is given in BS 3921 'Specification for clay bricks' as $225 \times 112.5 \times 75$ mm. This includes an allowance for a 10 mm mortar joint; thus the work size of the actual brick is $215 \times 102.5 \times 65$ mm.

Concrete bricks in accordance with BS 6073 Part 2 'Precast concrete masonry units' may be within any of the format ranges indicated in Table 4.1, which is based on BS 6073 Table 2.

Calcium silicate bricks in accordance with BS 187 'Specification for calcium silicate (sand lime and flint lime) bricks' have the same standard format as clay bricks.

Bricks can be classified in a number of ways with respect to their variety, type, quality and so on. However, for the purpose of this manual it will suffice to divide them into the following three general categories:

Facing bricks   These are clay or concrete bricks manufactured to satisfy aesthetic requirements. They are available in a wide range of strengths, colours and textures.

Table 4.1 Format range of concrete bricks (based on BS 6073 Part 2 1981 Table 2)

Work size of concrete bricks Length × thickness × height	Coordinating size of concrete bricks (including 10 mm mortar joints) Length × thickness × height
290 × 90 × 90	300 × 100 × 100
215 × 103 × 65	225 × 113 × 75
190 × 90 × 90	200 × 100 × 100
190 × 90 × 65	200 × 100 × 75

Common bricks   These are clay or concrete bricks produced for general building work and not to provide an attractive appearance. The term 'common' covers a wide variety of bricks and is not a guide to structural quality. Many common bricks have excellent strength properties.

Engineering bricks   These are clay bricks produced with defined compressive strength qualities. They are available in two classes: engineering A and engineering B.

4.4.2 Blocks

Blocks are walling units that exceed in length, width or height the sizes specified for bricks. They are generally produced from concrete.

In accordance with BS 6073 'Precast concrete masonry units' the purchaser of the blocks should specify their size from Table 1 in Part 2 of that standard, reproduced here as Table 4.2. To obtain the coordinating size of blockwork the nominal mortar joint, usually 10 mm, should be added to the length and height dimensions given in the table; the thickness remains unchanged. It should be noted that not every manufacturer will produce the complete range of work sizes given in the table.

Table 4.2 Work sizes of blocks (BS 6073 Part 2 1981 Table 1)

Length (mm)	Height (mm)	Thickness (mm)														
		60	75	90	100	115	125	140	150	175	190	200	215	220	225	250
390	190	×	×	×	×	×		×	×		×	×				
440	140	×	×	×	×			×	×		×	×			×	
440	190	×	×	×	×			×	×		×		×	×		
440	215	×	×	×	×	×	×	×	×	×	×	×	×	×	×	×
440	290	×	×	×	×			×	×		×	×	×			
590	140		×	×	×			×	×		×	×	×			
590	190		×	×	×			×	×		×	×	×			
590	215		×	×	×		×	×	×	×		×	×		×	×

The types of block generally available are as follows:

Facing blocks   Blocks with a finish suitable to provide an attractive appearance.

Ordinary or common blocks   Blocks suitable for internal use or, if rendered, for external use.

Solid blocks   These are primarily voidless, having no formal holes or cavities other than those inherent in the block material.

Hollow blocks   These are blocks which have cavities passing right through the unit, but the volume of such cavities must not exceed 50 per cent of the total unit volume.

Cellular blocks   These are similar to hollow blocks, but the cavities are effectively closed at one end. They are laid with the closed end uppermost in the wall to provide a good bed for the next layer of mortar.

Insulating blocks   These are usually cellular blocks faced with polystyrene or having the cavities filled with UF foam or polystyrene to improve their thermal qualities.

4.4.3 Mortar

Whilst masonry walls may be constructed from bricks, blocks or stone, in each of these the mortar is the common factor. The mortar serves several purposes in the construction, and must satisfy a number of requirements in both the newly mixed and the hardened state.

During construction, mortar should have good workability to enable efficient use by the bricklayer. It must spread easily so as to provide a level bed on which to align the masonry units of brick, block or stone. This in turn will ensure that the applied loads will be spread evenly over the bearing area of such units. When used with absorbent bricks it should retain moisture to avoid drying out and stiffening too quickly. Finally, it should harden in a reasonable time to prevent squeezing out under the pressure of the units laid above.

In the hardened state, mortar must be capable of transferring the stresses developed in the masonry units. Ideally, however, it should not be stronger than the masonry units themselves, so that any movement that occurs will be accommodated in the joints. This should ensure that any cracking that does develop will be in the mortar and not the masonry units.

Traditionally lime-sand mortars, relying on the loss of water and the action of carbonation to slowly gain strength, were employed for masonry construction. Whilst these offered excellent workability, their slow construction rate led to the adoption of cement mortars.

The addition of cement promotes a faster gain of strength, resulting in more rapid construction. Lime may still be included in the mix for workability, giving cement-lime-sand mortar. Ready mixed lime with sand may be obtained in specified proportions to which the cement is then added on site prior to use. Plasticized mortar is produced by replacing the lime with a proprietary plasticizer additive to provide the workability, giving a mix of cement and sand with plasticizer.

Mortar to which the cement has been added should generally be used within two hours of mixing. Ready mixed retarded mortars are available which contain a retarding agent to delay the set and prolong the working life of the mortar. These should not be used without the prior approval of the designer.

BS 5628 Part 1 Table 1 gives requirements for mortar designations in relation to their constituent proportions and compressive strength; this is reproduced here as Table 4.3. In general the lowest grade of mortar practicable should be used. Thus for general purpose masonry construction a 1:1:6 cement:lime:sand mortar will be sufficient. For high strength load bearing masonry a $1:\frac{1}{4}:3$ cement:lime:sand mortar is more appropriate. For reinforced masonry a mix not weaker than $1:\frac{1}{2}:4\frac{1}{2}$ cement:lime:sand should normally be specified.

The bond of the mortar with the masonry units is equally as important as its compressive strength. Adequate bond depends on a number of factors such as sand quality, the type and absorption rate of the masonry units at the time of laying, and attention to curing.

Ready mixed lime with sand for mortar should comply with the requirements of BS 4721 'Specification for ready mixed building mortars'. The mixing and use of mortars should be in accordance with the recommendations given in BS 5628 Part 3.

4.4.4   Wall ties

The two leaves of a cavity wall should be tied together by metal wall ties embedded at least 50 mm into the horizontal mortar joints. Their overall length should be chosen to suit the cavity width.

The ties should comply with the requirements of BS 1243 'Metal ties for cavity wall construction'. This code gives recommendations for three types of tie: the wire butterfly, the double triangle and the vertical twist. Ties can be manufactured from either galvanized or stainless steel.

The traditional butterfly tie has limited structural strength and is usually confined to domestic construction. Vertical twist wall ties are structurally the most substantial and are suitable for the most highly stressed load bearing cavity walls. Double triangle wall ties are less substantial than the vertical twist but better than the butterfly tie.

The minimum spacing and the selection of wall ties is dealt with in BS 5628 Part 3 Table 9, reproduced here as Table 4.4. Additional ties should be provided adjacent to wall openings in accordance with the recommendations given in the standard.

4.4.5   Damp proof courses

Whilst the main purpose of a damp proof course (DPC) is to provide a moisture barrier, in structural terms it must not squeeze out under vertical load or induce sliding under horizontal loading.

Table 4.3 Requirements for mortar (BS 5628 Part 1 1978 Table 1)

Properties	Mortar designation	Type of mortar (proportion by volume)			Mean compressive strength at 28 days (N/mm^2)	
		Cement:lime:sand	Masonry cement:sand	Cement:sand with plasticizer	Preliminary (laboratory) tests	Site tests
Increasing ability to accommodate movement, e.g. due to settlement, temperature and moisture changes	(i)	1:0 to $\frac{1}{4}$:3		—	16.0	11.0
	(ii)	1:$\frac{1}{2}$:4 to $4\frac{1}{2}$	1:$2\frac{1}{2}$ to $3\frac{1}{2}$	1:3 to 4	6.5	4.5
	(iii)	1:1:5 to 6	1:4 to 5	1:5 to 6	3.6	2.5
	(iv)	1:2:8 to 9	1:$5\frac{1}{2}$ to $6\frac{1}{2}$	1:7 to 8	1.5	1.0

Increasing strength →

Increasing resistance to frost attack during construction →

Improvement in bond and consequent resistance to rain penetration ←

Direction of change in properties is shown by the arrows

Table 4.4   Wall ties (BS 5628 Part 3 1985 Table 9)

(a) Spacing of ties

Least leaf thickness (one or both) (mm)	Type of tie	Cavity width (mm)	Equivalent no. of ties per square metre	Spacing of ties (mm) Horizontally	Vertically
65 to 90	All	50 to 75	4.9	450	450
90 or more	See table (b)	50 to 150	2.5	900	450

(b) Selection of ties

		Type of tie in BS 1243	Cavity width (mm)
↑ Increasing strength	Increasing flexibility and sound insulation ↓	Vertical twist	150 or less
		Double triangle	75 or less
		Butterfly	75 or less

A DPC can be made from a wide variety of materials, and therefore the choice should be based on the required performance in relation to the known behaviour of the materials. Advice on the physical properties and performance of DPC materials is given in BS 5628 Part 3.

4.5   Design philosophy

The design approach employed in BS 5628 is based on limit state philosophy. In the context of load bearing masonry its objective is to ensure an acceptable probability that the ultimate limit state will not be exceeded. Thus for a masonry member, which will be either a wall or a column,

$$\text{Ultimate design strength} \geq \text{ultimate design load}$$

4.6   Safety factors

As previously explained in relation to concrete design, partial safety factors are applied separately to both the loads and the material stresses in limit state design.

4.7   Loads

The basic or characteristic load is adjusted by a partial safety factor to arrive at the ultimate design load acting on a wall.

4.7.1   Characteristic loads

The characteristic loads applicable to masonry design are the same as those defined for concrete design:

Characteristic dead load G_k The weight of the structure complete with finishes, fixtures and partitions, obtained from BS 648 'Schedule of weights of building materials'.

Characteristic imposed load Q_k The live load produced by the occupants and usage of the building, obtained from BS 6399 'Design loading for buildings', Part 1 for floors or Part 3 for roofs.

Characteristic wind load W_k The wind load acting on the structure, obtained from CP 3 Chapter V Part 2 'Wind loads', eventually to become Part 2 of BS 6399.

4.7.2 Partial safety factors for load

As mentioned in relation to concrete design, the applied load may be greater in practice than the characteristic load for a number of reasons. To allow for such eventualities the respective characteristic loads are multiplied by a partial safety factor γ_f to give the ultimate design load appropriate to the load combination being considered. That is,

$$\text{Ultimate design load} = \gamma_f \times \text{characteristic load}$$

Values of γ_f are given in BS 5628 Part 1 for the following load combinations:

(a) Dead and imposed load
(b) Dead and wind load
(c) Dead, imposed and wind load
(d) Accidental damage.

Those for the dead and imposed load combination which would usually apply to vertically loaded walls are as follows:

Design dead load: $\gamma_f = 1.4 G_k$
Design imposed load: $\gamma_f = 1.6 Q_k$

4.7.3 Ultimate design load

The ultimate design load acting vertically on a wall will be the summation of the relevant characteristic load combinations multiplied by their respective partial safety factors. Therefore the ultimate design load for the dead plus imposed load combination on a vertically loaded wall would be expressed as follows:

$$\text{Ultimate design load dead} + \text{imposed} = \gamma_f G_k + \gamma_f Q_k = 1.4 G_k + 1.6 Q_k$$

4.8 Material properties

Like concrete, the strength of masonry materials in an actual wall can differ from their specified strength for a number of reasons. The characteristic strength f_k of the masonry units is therefore divided by a partial safety factor γ_m to arrive at the ultimate design strength of the units. In relation to vertically loaded walls it is the compressive strength we are usually concerned with.

4.8.1 Characteristic compressive strength of masonry units

The characteristic compressive strength f_k for various masonry units is given in BS 5628 Part 1 Table 2a–d, reproduced here as Table 4.5a–d. It depends on the basic compressive strength of particular masonry units in conjunction with the designated mortar mix.

Table 4.5 Characteristic compressive strength of masonry f_k (N/mm^2) (BS 5628 Part 1 1978 Table 2)

INTERPOLATE FOR f_k.

(a) Constructed with standard format bricks

Mortar designation	Compressive strength of unit (N/mm²)								
	5	10	15	20	27.5	35	50	70	100
(i)	2.5	4.4	6.0	7.4	9.2	11.4	15.0	19.2	24.0
(ii)	2.5	4.2	5.3	6.4	7.9	9.4	12.2	15.1	18.2
(iii)	2.5	4.1	5.0	5.8	7.1	8.5	10.6	13.1	15.5
(iv)	2.2	3.5	4.4	5.2	6.2	7.3	9.0	10.8	12.7

(b) Constructed with blocks having a ratio of height to least horizontal dimension of 0.6

Mortar designation	Compressive strength of unit (N/mm²)							
	2.8	3.5	5.0	7.0	10	15	20	35 or greater
(i)	1.4	1.7	2.5	3.4	4.4	6.0	7.4	11.4
(ii)	1.4	1.7	2.5	3.2	4.2	5.3	6.4	9.4
(iii)	1.4	1.7	2.5	3.2	4.1	5.0	5.8	8.5
(iv)	1.4	1.7	2.2	2.8	3.5	4.4	5.2	7.3

(c) Constructed with hollow blocks having a ratio of height to least horizontal dimension of between 2.0 and 4.0

Mortar designation	Compressive strength of unit (N/mm²)							
	2.8	3.5	5.0	7.0	10	15	20	35 or greater
(i)	2.8	3.5	5.0	5.7	6.1	6.8	7.5	11.4
(ii)	2.8	3.5	5.0	5.5	5.7	6.1	6.5	9.4
(iii)	2.8	3.5	5.0	5.4	5.5	5.7	5.9	8.5
(iv)	2.8	3.5	4.4	4.8	4.9	5.1	5.3	7.3

(d) Constructed from solid concrete blocks having a ratio of height to least horizontal dimension of between 2.0 and 4.0

Mortar designation	Compressive strength of unit (N/mm²)							
	2.8	3.5	5.0	7.0	10	15	20	35 or greater
(i)	2.8	3.5	5.0	6.8	8.8	12.0	14.8	22.8
(ii)	2.8	3.5	5.0	6.4	8.4	10.6	12.8	18.8
(iii)	2.8	3.5	5.0	6.4	8.2	10.0	11.6	17.0
(iv)	2.8	3.5	4.4	5.6	7.0	8.8	10.4	14.6

The basic compressive strength of the individual masonry units given in each part of the table is based upon tests which take into account the presence of any voids or perforations in the unit. Thus the structural calculations for a wall constructed from either solid or hollow units can be made in exactly the same way.

The designation of mortar types is given in BS 5628 Part 1 Table 1, reproduced earlier as Table 4.3.

To obtain the respective value of f_k, reference should be made to the relevant part of Table 4.5 as explained in the following sections.

Bricks

Generally for bricks of standard dimensional format, f_k is obtained directly from Table 4.5a.

However, if a solid wall or the loaded inner leaf of a cavity wall is constructed with standard format bricks, and the wall or leaf thickness is equal to the width of a single brick, then the value of f_k from Table 4.5a may be multiplied by 1.15. This increase in the compressive strength is based upon tests which have shown that such walls are stronger owing to the absence of vertical mortar joints within the wall thickness. It should be noted that this factor of 1.15 does not apply to cavity walls where both leaves are loaded.

Blocks

When a wall is constructed with blockwork, the increased size of the individual masonry units means that there are fewer joints compared with an equivalent wall of standard format bricks. Fewer joints result in a stronger wall, and hence the characteristic compressive strength of blockwork is influenced by the shape of the individual units.

The shape factor of a block is obtained by dividing its height by its lesser horizontal dimension. For example, for the block shown in Figure 4.2,

$$\text{Shape factor} = \frac{\text{height}}{\text{lesser horizontal dimension}} = \frac{200}{100} = 2$$

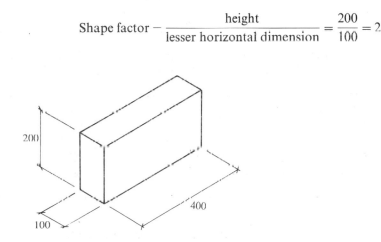

Figure 4.2 *Dimensions of a typical block*

Depending on the shape factor and the type of block, f_k is then obtained from the relevant part of Table 4.5:

(a) For hollow and solid blocks having a shape factor not greater than 0.6, f_k is obtained directly from Table 4.5b.

(b) For hollow blocks having a shape factor between 2.0 and 4.0, f_k is obtained directly from Table 4.5c.

(c) For solid blocks having a shape factor between 2.0 and 4.0, f_k is obtained directly from Table 4.5d.

In certain circumstances interpolation between the tables may be necessary as follows:

(d) For hollow block walls having a shape factor between 0.6 and 2.0, f_k is obtained by interpolation between the values in Table 4.5b and Table 4.5c.

(e) For solid block walls having a shape factor between 0.6 and 2.0, f_k is obtained by interpolation between the values in Table 4.5b and 4.5d.

Natural stone

For natural stone, f_k should generally be taken as that for solid concrete blocks of equivalent strength.

Random rubble masonry

For random rubble masonry f_k should be taken as 75 per cent of that for the corresponding strength of natural stone.

Modification to characteristic strength for shell bedding

Hollow concrete blocks are sometimes laid on a mortar bed consisting of two strips along the outer edges of the block. This is termed 'shell bedding' and is illustrated in Figure 4.3.

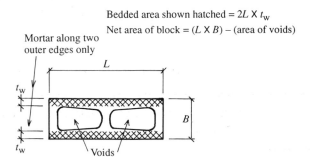

Bedded area shown hatched = $2L \times t_w$

Net area of block = $(L \times B)$ – (area of voids)

Mortar along two outer edges only

t_w

L

t_w

B

Voids

Figure 4.3 *Shell bedding to hollow blocks*

If such a construction procedure is to be permitted then the design calculations should be adjusted accordingly by reducing the characteristic strength. This is done by multiplying the value of f_k obtained from either Table 4.5b or Table 4.5c by a factor equal to the bedded area divided by the net area of the block:

$$\text{Shell bedded } f_k = f_k \text{ from table} \times \frac{\text{bedded area}}{\text{net area of block}}$$

Modification to characteristic strength for small plan areas

When the horizontal cross-sectional area of a loaded wall or column is less than $0.2\,\text{m}^2$, the value of f_k obtained from the tables should be multiplied by the following modification factor:

$$\text{Small plan area modification factor} = 0.7 + 1.5A$$

where A is the loaded horizontal cross-sectional area of the wall or column (m^2).

4.8.2 Partial safety factors for materials γ_m

The partial safety factor γ_m for materials in masonry design is obtained from BS 5628 Part 1 Table 4, reproduced here as Table 4.6.

Table 4.6 Partial safety factors for material strength γ_m (BS 5628 Part 1 1978 Table 4)

Category of manufacturing control of structural units	Category of construction control	
	Special	Normal
Special	2.5	3.1 BRick
Normal	2.8	3.5 Black

The factor is related to the standard of quality control exercised during both the manufacture and construction stages. In each case two levels of control are recognized, normal category or special category, and these apply as follows.

Normal category of manufacturing control

This should be assumed when the materials to be supplied will simply comply with the compressive strength requirements of the relevant British Standard.

Special category of manufacturing control

This may be assumed when the manufacturer agrees to supply materials that comply with a specified strength limit. Furthermore, the supplier must operate a quality control system to provide evidence that such a limit is being consistently met.

Normal category of construction control

This should be assumed when the standard of workmanship is in accordance with the recommendations given in BS 5628 Part 3, and appropriate site supervision and inspection will be carried out to ensure that this is so. Some of the construction aspects covered by these workmanship requirements are as follows:

(a) Setting out
(b) Storage of materials
(c) Batching, mixing and use of mortars
(d) Laying of masonry units
(e) Constructional details
(f) Protection during construction.

Special category of construction control

This may be assumed when, in addition to the normal category requirements, compliance testing of the mortar strength will be carried out in accordance with Appendix A of BS 5628 Part 1.

4.8.3 Ultimate compressive strength of masonry units

The ultimate compressive strength of masonry units, as mentioned earlier, is obtained by dividing the characteristic strength by the appropriate partial safety factor:

$$\text{Ultimate compressive strength} = \frac{\text{characteristic strength of units}}{\text{partial safety factor}} = \frac{f_k}{\gamma_m}$$

Having arrived at an ultimate compressive strength for the masonry units that are to be used, the next step is to determine the load bearing capacity of the particular member in which they are to be incorporated. In terms of masonry design such members will either be walls or columns.

4.9 Factors influencing the load bearing capacity of masonry members

There are a number of interrelated factors that influence the load bearing capacity of masonry walls and columns:

(a) Slenderness ratio
(b) Lateral support
(c) Effective height h_{ef}
(d) Effective length l_{ef}
(e) Effective thickness t_{ef}
(f) Capacity reduction factor for slenderness.

The principal factor is the slenderness ratio; all the others are related to it. Let us therefore consider the effect of each factor on walls and columns.

4.9.1 Slenderness ratio

Vertically loaded walls and columns can fail by crushing due to direct compression or, if they are slender, by lateral buckling. A measure of the tendency to fail by buckling before crushing is the slenderness ratio (SR).

In accordance with BS 5628 the slenderness ratio of a wall should be calculated as follows:

$$\text{SR wall} = \frac{\text{effective height}}{\text{effective thickness}} \quad \text{or} \quad \frac{\text{effective length}}{\text{effective thickness}}$$

$$= \frac{h_{ef}}{t_{ef}} \quad \text{or} \quad \frac{l_{ef}}{t_{ef}}$$

The effective length is only used when this would give a lesser slenderness ratio value.

For masonry columns the effective height is always used when calculating the slenderness ratio:

$$\text{SR column} = \frac{\text{effective height}}{\text{effective thickness}} = \frac{h_{ef}}{t_{ef}}$$

The slenderness ratio of a member should generally not exceed 27. However, should the thickness of a wall be less than 90 mm, in a building of two storeys, then the slenderness ratio value must not exceed 20.

4.9.2 Lateral support

The effective height and the effective length are influenced by the degree of any lateral support that may be provided. With respect to the height this will be provided in the horizontal direction by the floors or roof. In the case of the length it will be provided in the vertical direction by any intersecting or return walls.

BS 5628 defines the degree of resistance to lateral movement as either 'simple' or 'enhanced' depending on the construction details adopted. Examples of horizontal lateral support that only provide simple resistance are illustrated in Figure 4.4; those capable of providing enhanced resis-

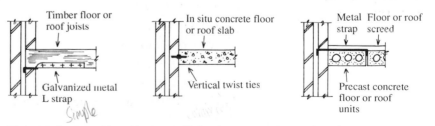

Figure 4.4 *Examples of horizontal lateral support only capable of providing simple resistance*

tance are illustrated in Figure 4.5. Similarly, examples of vertical lateral support that only provide simple resistance are shown in Figure 4.6; those that provide enhanced resistance are shown in Figure 4.7.

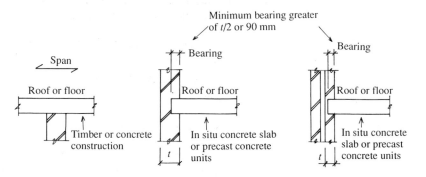

Figure 4.5 *Examples of horizontal lateral support capable of providing enhanced resistance*

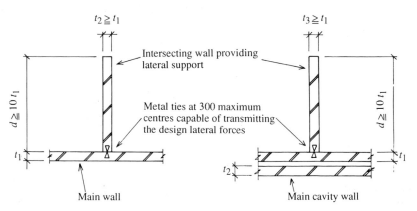

Figure 4.6 *Examples of vertical lateral support only capable of providing simple resistance*

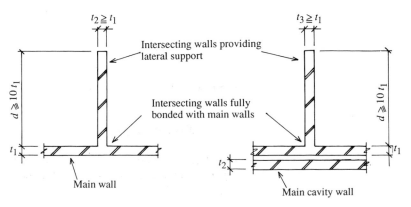

Figure 4.7 *Examples of vertical lateral support capable of providing enhanced resistance*

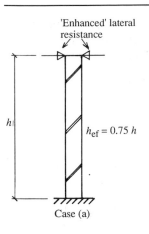

'Enhanced' lateral resistance

$h_{ef} = 0.75\,h$

Case (a)

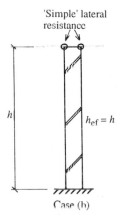

'Simple' lateral resistance

$h_{ef} = h$

Case (b)

Figure 4.8 *Effective height of walls*

4.9.3 Effective height

The effective height h_{ef} depends on the degree of horizontal lateral support provided and may be defined as follows for walls and columns.
  For walls it should be taken as

(a) 0.75 times the clear distance between lateral supports which provide enhanced resistance, as depicted in Figure 4.8a; or

(b) The clear distance between lateral supports which only provide simple resistance, as depicted in Figure 4.8b.

For columns it should be taken as

(a) The distance between lateral supports in respect of the direction in which lateral support is provided, shown as $h_{ef} = h$ in Figure 4.9a and b; or

(b) Twice the height of the column in respect of a direction in which lateral support is not provided, shown as $h_{ef} = 2h$ in Figure 4.9b.

It should be noted that BS 5628 suggests that lateral support to columns should preferably be provided in both horizontal directions.

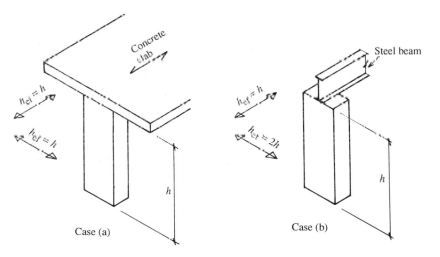

Case (a)   Case (b)

Figure 4.9 *Effective height of columns*

4.9.4 Effective length

The effective length l_{ef} is a consideration that only applies to walls, and depends on the degree of vertical lateral support provided. It may be taken as

(a) 0.75 times the clear distance between lateral supports which provide enhanced resistance, as illustrated in Figure 4.10a

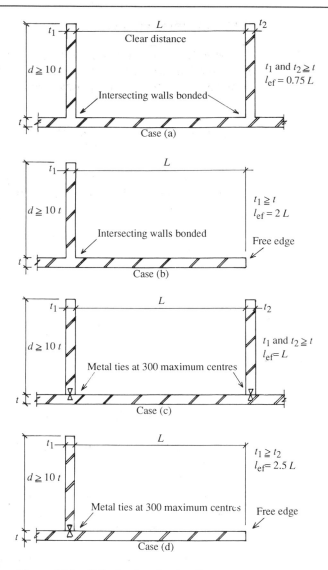

Figure 4.10   *Effective length of walls*

(b) Twice the distance between a lateral support which provides enhanced resistance and a free edge, as illustrated in Figure 4.10b

(c) The clear distance between lateral supports which only provided simple resistance, as illustrated in Figure 4.10c

(d) 2.5 times the distance between a lateral support which provides simple resistance and a free edge, as illustrated in Figure 4.10d.

It should be appreciated that the slenderness ratio of a wall without any vertical lateral supports must be based upon its effective height.

4.9.5 Effective thickness

The effective thickness t_{ef} parameters for walls and columns are illustrated in Figure 2 of BS 5628. They are basically divided into two categories in relation to whether stiffening piers or intersecting walls are present or not.

Category 1   walls and columns not stiffened by piers or intersecting walls

(a) Columns as shown in Figure 4.11: $t_{ef} = t$ or b depending in which direction the slenderness is being considered.

(b) Single leaf walls as shown in Figure 4.12: $t_{ef} =$ the actual thickness t.

(c) Cavity walls as shown in Figure 4.13: $t_{ef} =$ the greatest of $2(t_1 + t_2)/3$ or t_1 or t_2.

$b \not> 4t$

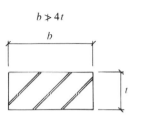

Figure 4.11   *Plan on a column*

Figure 4.12   *Plan on a single leaf wall*

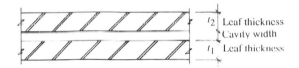

Figure 4.13   *Plan on a cavity wall*

Category 2:   walls stiffened by piers or intersecting walls

(a) Single leaf wall with piers shown in Figure 4.14: $t_{ef} = tK$, where K is the appropriate stiffness coefficient from BS 5628 Table 5, reproduced here as Table 4.7.

(b) Cavity wall with piers as shown in Figure 4.15: $t_{ef} =$ the greatest of $2(t_1 + Kt_2)/3$ or t_1 or Kt_2, where K is again the appropriate stiffness coefficient from Table 4.7.

For the purpose of category 2 an intersecting wall may be assumed to be equivalent to a pier with the dimensions shown in Figure 4.16.

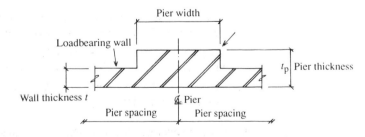

Figure 4.14   *Plan on a single leaf wall with piers*

Table 4.7   Stiffness coefficient for walls stiffened by piers (BS 5628 Part 1 1978 Table 5)

Ratio of pier spacing (centre to centre) to pier width	Ratio t_p/t of pier thickness to actual thickness of wall to which it is bonded		
	1	2	3
6	1.0	1.4	2.0
10	1.0	1.2	1.4
20	1.0	1.0	1.0

Note: Linear interpolation between the values given in table is permissible, but not extrapolation outside the limits given.

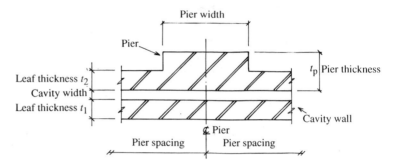

Figure 4.15   *Plan on a cavity wall with piers*

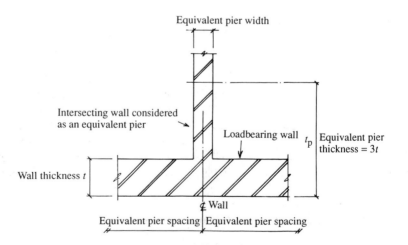

Figure 4.16   *Plan on an intersecting wall considered as an equivalent pier*

4.9.6   Capacity reduction factor for slenderness

As stated earlier, the slenderness ratio is a measure of the tendency of a wall or column to fail by buckling before crushing. To take this into account, the design strength of a wall or column is reduced using a capacity reduction factor β which is based upon the slenderness ratio value. It is obtained from BS 5628 Part 1 Table 7, reproduced here as Table 4.8.

A load applied eccentrically will increase the tendency for a wall or column to buckle and reduce the load capacity further. This is catered for by using a modified capacity reduction factor β from Table 4.8 which depends on the ratio of the eccentricity e_x to the member thickness.

Table 4.8 Capacity reduction factor β (BS 5628 Part 1 1978 Table 7)

Slenderness ratio h_{ef}/t_{ef}	Eccentricity at top of wall e_x			
	Up to $0.05t$ (see note 1)	$0.1t$	$0.2t$	$0.3t$
0	1.00	0.88	0.66	0.44
6	1.00	0.88	0.66	0.44
8	1.00	0.88	0.66	0.44
10	0.97	0.88	0.66	0.44
12	0.93	0.87	0.66	0.44
14	0.89	0.83	0.66	0.44
16	0.83	0.77	0.64	0.44
18	0.77	0.70	0.57	0.44
20	0.70	0.64	0.51	0.37
22	0.62	0.56	0.43	0.30
24	0.53	0.47	0.34	
26	0.45	0.38		
27	0.40	0.33		

Note 1: It is not necessary to consider the effects of eccentricities up to and including $0.05t$.
Note 2: Linear interpolation between eccentricities and slenderness ratios is permitted.
Note 3: The derivation of β is given in Appendix B of BS 5628.

Whilst ideally the actual eccentricity should be calculated, BS 5628 allows it to be assumed at the discretion of the designer. Thus for a wall supporting a single floor or roof it may be assumed that the load will act at one-third of the bearing length from the edge of the supporting wall or leaf, as illustrated in Figure 4.17. When a floor of uniform thickness is continuous over a supporting wall, each span of the floor may be taken as being supported on half the total bearing area, as shown in Figure 4.18.

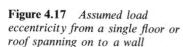

Figure 4.17 *Assumed load eccentricity from a single floor or roof spanning on to a wall*

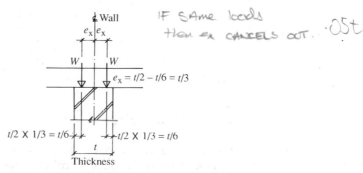

Figure 4.18 *Assumed load eccentricity from a floor or roof continuous over a wall*

Where joist hangers are used the resultant load should be assumed to be applied at a distance of 25 mm from the face of the wall, as shown in Figure 4.19.

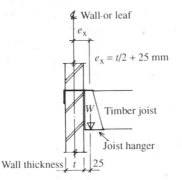

Figure 4.19   *Assumed load eccentricity from joist hangers*

4.10   Vertical load resistance

The design vertical resistance of a wall per unit length is given by the following expression:

$$\text{Vertical design strength per unit length of wall} = \frac{\beta t f_k}{\gamma_m} \quad (4.1)$$

where

- β capacity reduction factor from Table 4.8
- f_k characteristic strength of masonry units from the appropriate part of Table 4.5
- γ_m material partial safety factor from Table 4.6
- t actual thickness of leaf or wall

For a rectangular masonry column the design vertical load resistance is given by the following expression:

$$\text{Vertical design strength of column} = \frac{\beta b t f_k}{\gamma_m} \quad (4.2)$$

where b is the width of the column, t is the thickness of the column, and the other symbols are the same as those defined after expression 4.1 for walls.

The design vertical load resistance of a cavity wall or column is determined in relation to how the vertical load is applied. When the load acts on the centroid of the two leaves (Figure 4.20a) it should be replaced by two statically equivalent axial loads acting on each of the leaves (Figure 4.20b). Each leaf should then be designed to resist the equivalent

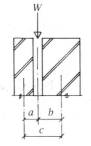

(a) Load acting on centroid of cavity wall

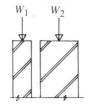

$W_1 = Wb/c$ and $W_2 = Wa/c$

(b) Equivalent axial load acting on each leaf

Figure 4.20   *Cavity wall with both leaves loaded*

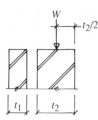

Figure 4.21 *Cavity wall with only one leaf loaded*

axial load it supports using the appropriate expression 4.1 or 4.2; the effective thickness of the wall for the purpose of obtaining the capacity reduction factor from Table 4.8 is that of the cavity wall or column.

The load from a roof or floor is often only supported by one leaf of a cavity wall, as shown in Figure 4.21. Then the design strength should be calculated using the thickness of that leaf alone in the relevant expression 4.1 or 4.2. The effective thickness used for obtaining the capacity reduction factor is again that of the cavity wall, thus taking into account the stiffening effect of the other leaf.

The general procedure for determining the vertical design strength of a wall or column may be summarized as follows.

4.10.1 Design summary for a vertically loaded wall or column

(a) Calculate the slenderness ratio for the wall or column under consideration.

(b) Obtain the capacity reduction factor β from Table 4.8 corresponding to the slenderness ratio and taking into account any eccentricity of loading.

(c) Obtain the characteristic compressive strength f_k of the masonry units from the relevant part of Table 4.5, adjusting if necessary for the plan area or shell bedding.

(d) Select the material partial safety factor γ_m from Table 4.6 in relation to the standard of quality control that will be exercised.

(e) Calculate the vertical load resistance using expression 4.1 for walls or expression 4.2 for columns.

Whilst following this procedure, particular care needs to be exercised by the designer to ensure that all the factors that can influence the slenderness ratio are taken into consideration. Let us therefore look at a number of examples using this procedure which attempt to highlight those various factors.

Example 4.1

A 102.5 mm thick single skin brick wall, as shown in Figure 4.22, is built between the concrete floors of a multi-storey building. It supports an ultimate axial load,

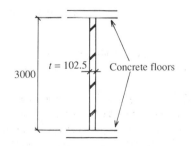

Figure 4.22 *Section through wall*

including an allowance for the self-weight, of 250 kN per metre run. What brick and mortar strengths are required if normal manufacturing and construction controls apply and the wall is first 10 m long and secondly only 1 m long?

Wall 10 m long

Since the wall in this instance is not provided with any vertical lateral supports along its 10 m length, the slenderness ratio should be based upon its effective height. Furthermore, as the concrete floor is continuous over the wall, by reference to Figure 4.5 enhanced lateral resistance will be provided in the horizontal direction.

The effective height of the wall $h_{ef} = 0.75h = 0.75 \times 3000 = 2250$ mm, and the effective thickness of a solid wall is the actual thickness of 102.5 mm. Thus the slenderness ratio is given by

$$\text{SR} = \frac{\text{effective height}}{\text{effective thickness}} = \frac{h_{ef}}{t_{ef}} = \frac{2250}{102.5} = 21.95 < 27$$

Thus the slenderness ratio is acceptable. From Table 4.8, the capacity reduction factor β is 0.62.

Since the wall is 10 m long, the plan area is $10 \times 0.1025 = 1.025\,\text{m}^2$. This is greater than $0.2\,\text{m}^2$ and therefore the plan area reduction factor does not apply.

Now the ultimate vertical load is 250 kN per metre run or 250 N per millimetre run. The expression for the vertical design strength of a wall is $\beta t f_k / \gamma_m$. Therefore

$$250\,\text{N per mm run} = \frac{\beta t f_k}{\gamma_m}$$

The material partial safety factor γ_m is selected from Table 4.6 in relation to the standard of manufacture and construction control. In this instance it is normal for both manufacture and construction, and γ_m will therefore be 3.5. Furthermore, since the thickness of the wall is equal to the width of a single brick, the value of f_k may be multiplied by 1.15. Hence

$$250 = \frac{\beta t f_k \times 1.15}{\gamma_m}$$

from which

$$f_k \text{ required} = \frac{250 \gamma_m}{\beta t \times 1.15} = \frac{250 \times 3.5}{0.62 \times 102.5 \times 1.15} = 11.97\,\text{N/mm}^2$$

Comparing this value with the characteristic strength of bricks given in Table 4.5a, suitable bricks and mortar can be chosen:

Use 50 N/mm² bricks in grade (ii) mortar ($f_k = 12.2\,\text{N/mm}^2$).   TAB 4.5.

Wall 1 m long

In this instance the plan area of the wall is $1 \times 0.1025 = 0.1025\,\text{m}^2$. This is less than $0.2\,\text{m}^2$, and hence the small plan area modification factor should be applied:

Modification factor $= (0.7 + 1.5A) = (0.7 + 1.5 \times 0.1025) = 0.854$

The characteristic strength f_k obtained from Table 4.5a should be multiplied by this factor in addition to the single skin factor of 1.15. Thus

$$250 \text{ N per mm run} = \frac{\beta t f_k \times 1.15 \times 0.854}{\gamma_m}$$

from which

$$f_k \text{ required} = \frac{250 \gamma_m}{\beta t \times 1.15 \times 0.854} = \frac{250 \times 3.5}{0.62 \times 102.5 \times 1.15 \times 0.854}$$

$$= 14.02 \text{ N/mm}^2$$

Again by reference to Table 4.5a:

Use 50 N/mm^2 bricks in grade (i) mortar ($f_k = 15 \text{ N/mm}^2$), or use 70 N/mm^2 bricks in grade (ii) mortar ($f_k = 15.1 \text{ N/mm}^2$).

Example 4.2

A single skin wall constructed from 390 mm long × 190 mm high × 100 mm thick solid concrete blocks is built between concrete floors as shown in Figure 4.23. The ultimate axial load carried by the wall, including an allowance for the self-weight, is 125 kN per metre run. If the wall is 5 m long, what block and mortar strengths are required if special manufacturing control and normal construction control will apply?

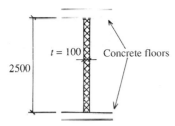

$t = 100$ Concrete floors

2500

Figure 4.23   *Section through wall*

Since there are no intersecting walls the effective height will govern the slenderness. The effective height $h_{ef} = 0.75h = 0.75 \times 2500 = 1875$ mm, and the effective thickness t_{ef} is the actual thickness of 100 mm. Thus

$$SR = \frac{h_{ef}}{t_{ef}} = \frac{1875}{100} = 18.75 < 27$$

Thus the slenderness ratio is acceptable. By interpolation from Table 4.8, the capacity reduction factor β is 0.74.

The plan area of the 5 m long wall is $5 \times 0.1 = 0.5 \text{ m}^2$. This is greater than 0.2 m^2 and therefore the plan area reduction factor does not apply. Furthermore, it should be appreciated that the single skin factor used for the brick wall in Example 4.1 does not apply to walls constructed from blocks.

The vertical design strength is $\beta t f_k / \gamma_m$. Thus

$$125 = \frac{\beta t f_k}{\gamma_m}$$

from which

$$f_k \text{ required} = \frac{125 \gamma_m}{\beta t} = \frac{125 \times 3.1}{0.74 \times 100} = 5.24 \text{ N/mm}^2$$

Now the blocks to be used are solid concrete 390 mm long × 190 mm high × 100 mm thick, for which the ratio of the block height to the lesser horizontal dimension is $390/100 = 3.9$. Therefore f_k should be obtained from Table 4.5d:

Use 7 N/mm^2 solid blocks in grade (iv) mortar ($f_k = 5.6 \text{ N/mm}^2$).

Example 4.3

A grou..d floor wall in a three-storey building supports the loads indicated in Figure 4.24. Choose suitable bricks and mortar for the wall. Partial safety factors are given as follows: for materials, $\gamma_m = 2.8$; for dead loads, $\gamma_f = 1.4$; for imposed loads, $\gamma_f = 1.6$. The manufacturing control is to be normal and the construction control is to be special.

Floor characteristic dead load = 4.8 kN/m^2
Floor characteristic imposed load = 5 kN/m^2

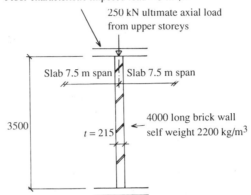

Figure 4.24   *Section through wall*

Consider a 1 m length of wall for the purpose of design.

The ultimate design load from the upper storey has been given, but to this must be added the first-floor loading and the self-weight of the ground floor wall itself. Hence the characteristic dead load G_k is calculated as follows:

Floor: $4.8 \times 7.5 \times 1 =$ 36

Wall SW: $\dfrac{2000}{100} \times 3.5 \times 0.215 \times 1 = 16.56$

Total G_k: $\overline{52.56}$ kN

The characteristic imposed load will be $Q_k = 5 \times 7.5 \times 1 = 37.5$ kN. Hence

Ultimate design dead and imposed load

$$= \gamma_f G_k + \gamma_f Q_k = 1.4 \times 52.56 + 1.6 \times 37.5 = 73.58 + 60 = 133.58 \text{ kN per metre run}$$

To this must be added the ultimate design load from the upper storeys of 250 kN per metre run. Hence

Total ultimate axial load $= 250 + 133.58 = 383.58$ kN/m $= 383.58$ N per mm run

The effective height $h_{ef} = 0.75h = 0.75 \times 3500 = 2625$ mm, and the effective thickness t_{ef} is the actual thickness of 215 mm. Note that since the thickness of this brick wall is greater than a standard format brick, the thickness factor 1.15 does not apply. Thus the slenderness ratio is given by

$$\text{SR} = \frac{h_{ef}}{t_{ef}} = \frac{2625}{215} = 12.2 < 27$$

This is satisfactory. From Table 4.8, the capacity reduction factor β is 0.93.

The plan area of the 4 m long wall is $4 \times 0.215 = 0.86\,\text{m}^2$. This is greater than $0.2\,\text{m}^2$, and therefore the plan area reduction factor does not apply.

The expression for the vertical design strength per unit length of walls is $\beta t f_k / \gamma_m$. Therefore

$$383.58 = \frac{\beta t f_k}{\gamma_m}$$

from which

$$f_k \text{ required} = \frac{383.58 \gamma_m}{\beta t} = \frac{383.58 \times 2.8}{0.93 \times 215} = 5.37\,\text{N/mm}^2$$

By reference to Table 4.5a:

Use $20\,\text{N/mm}^2$ bricks in grade (iii) mortar ($f_k = 5.8\,\text{N/mm}^2$).

Example 4.4

The brick cavity wall shown in Figure 4.25 supports an ultimate load on the inner leaf of 75 kN/m, the outer leaf being unloaded. Select suitable bricks and mortar if both the manufacturing and construction control are to be normal.

The effective height $h_{ef} = 0.75h = 0.75 \times 4000 = 3000\,\text{mm}$. The effective thickness t_{ef} is the greatest of $2(t_1 + t_2)/3 = 2(102.5 + 102.5)/3 = 136.7\,\text{mm}$, or $t_1 = 102.5\,\text{mm}$, or $t_2 = 102.5\,\text{mm}$. Thus the slenderness ratio is given by

$$SR = \frac{h_{ef}}{t_{ef}} = \frac{3000}{136.7} = 21.95 < 27$$

This is satisfactory. The load from the roof slab will be applied eccentrically as shown in Figure 4.26; that is, the eccentricity is given by

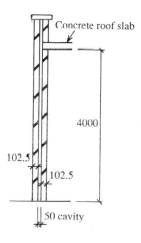

Figure 4.25 *Section through wall*

$$e_x = \frac{t}{2} - \frac{t}{3} = \frac{t}{6} = 0.167t$$

Hence from Table 4.8 the capacity reduction factor β is 0.473.

The vertical design strength per unit length of wall is $\beta t f_k / \gamma_m$. Therefore

$$75\,\text{N/mm} = \frac{\beta t f_k \times 1.15}{\gamma_m}$$

from which

$$f_k \text{ required} = \frac{75 \gamma_m}{\beta t \times 1.15} = \frac{75 \times 3.5}{0.473 \times 102.5 \times 1.15} = 4.7\,\text{N/mm}^2$$

Figure 4.26 *Load eccentricity*

By reference to Table 4.5a:

Use $15\,\text{N/mm}^2$ bricks in grade (iii) mortar ($f_k = 5\,\text{N/mm}^2$).

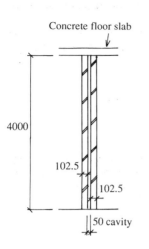

Concrete floor slab

4000

102.5

102.5

50 cavity

Figure 4.27 *Section through wall*

Example 4.5

The brick cavity wall shown in Figure 4.27 supports an ultimate axial load of 150 kN/m shared equally by both leaves. Select suitable bricks and mortar if both the manufacturing and construction control are to be normal.

The effective height and thickness and hence the slenderness ratio are the same as in Example 4.4; that is, SR = 21.95. However in this example, since the two leaves of the wall share the load equally, there is no eccentricity. Hence from Table 4.8 the capacity reduction factor β is 0.62.

The vertical design strength is $\beta t f_k / \gamma_m$. Thus, for each leaf,

$$f_k \text{ required} = \frac{150 \gamma_m}{\beta t} = \frac{150 \times 3.5}{0.62 \times 2 \times 102.5} = 4.13 \text{ N/mm}^2$$

It should be noted that the narrow brick wall factor of 1.15 does not apply in this instance since both leaves are loaded. From Table 4.5a:

Use 15 N/mm² bricks in grade (iv) mortar ($f_k = 4.4$ N/mm²).

Example 4.6

The wall shown in Figure 4.28 is built of 50 N/mm² clay bricks set in grade (i) mortar. Calculate the vertical design strength of the wall if it is 2.4 m high and is provided with simple lateral support at the top. The category of manufacturing control is to be normal and that for construction special.

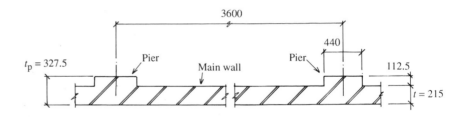

3600

440

$t_p = 327.5$ Pier   Pier   112.5

Main wall

$t = 215$

Figure 4.28 *Plan on wall*

The effective height with simple lateral resistance is $h_{ef} = h = 2400$ mm. Since vertical lateral support is not provided, the effective height will govern the slenderness. The effective thickness will be influenced by the piers:

$$\frac{\text{Pier spacing}}{\text{Pier width}} = \frac{3600}{440} = 8.18$$

$$\frac{\text{Pier thickness}}{\text{Wall thickness}} = \frac{t_p}{t} = \frac{327.5}{215} = 1.52$$

Therefore, by interpolation from Table 4.7, the stiffness coefficient $K = 1.151$. Hence the effective thickness $t_{ef} = tK = 215 \times 1.151 = 247.47$ mm. Thus the slenderness ratio is given by

$$\text{SR} = \frac{h_{ef}}{t_{ef}} = \frac{2400}{247.47} = 9.7 < 27$$

This is satisfactory. Hence by interpolation from Table 4.8 the capacity reduction factor β is 0.975 without eccentricity.

From Table 4.5a, the masonry characteristic strength $f_k = 15\,\text{N/mm}^2$. The material partial safety factor $\gamma_m = 2.8$. Finally, the ultimate vertical design strength per unit length of wall is

$$\frac{\beta t f_k}{\gamma_m} = \frac{0.975 \times 215 \times 15}{2.8} = 1122.99\,\text{N/mm} = 1122.99\,\text{kN per metre run}$$

Example 4.7

Calculate the vertical design strength of the wall shown in Figure 4.29, assuming simple lateral support is provided at the top. The wall is 3.45 m high and is constructed from 27.5 N/mm² bricks set in grade (iii) mortar, and both the manufacturing and construction control are normal.

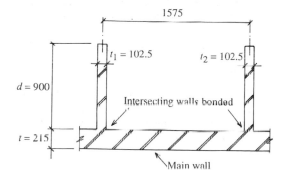

Figure 4.29 *Plan on wall*

The effective height $h_{ef} = h = 3450\,\text{mm}$. The intersecting walls are not long enough (that is $d < 10t$) or thick enough (that is t_1 and $t_1 < t$) to provide enhanced lateral support in the vertical direction; therefore the effective height will govern the slenderness. However, the length of the intersecting walls is greater than $3t$ and they may therefore be considered as equivalent stiffening piers. That is,

$$\frac{\text{Equivalent pier spacing}}{\text{Equivalent pier width}} = \frac{1575}{102.5} = 15.37$$

Equivalent pier thickness $t_p = 3t = 3 \times 215 = 645\,\text{mm}$

$$\frac{\text{Equivalent pier thickness}}{\text{Wall thickness}} = \frac{t_p}{t} = \frac{645}{215} = 3$$

Therefore, by interpolation from Table 4.7, the stiffness coefficient K is 1.19. The effective thickness $t_{ef} = tK = 215 \times 1.19 = 255.85\,\text{mm}$. Thus the slenderness ratio is given by

$$\text{SR} = \frac{h_{ef}}{t_{ef}} = \frac{3450}{255.85} = 13.48 < 27$$

This is satisfactory. By interpolation from Table 4.7, the capacity reduction factor β is 0.90 without eccentricity.

From Table 4.5a, the masonry characteristic strength $f_k = 7.1\,\text{N/mm}^2$. The material partial safety factor $\gamma_m = 3.5$. Thus the ultimate vertical design strength is

$$\frac{\beta t f_k}{\gamma_m} = \frac{0.9 \times 215 \times 7.1}{3.5} = 392.53\,\text{N/mm} = 392.53\,\text{kN per metre run}$$

Example 4.8

Determine the vertical design strength of the wall shown in Figure 4.30. The wall is 3.45 m high, restrained at the top, and constructed from $35\,\text{N/mm}^2$ bricks set in grade (iii) mortar. Both the manufacturing and construction control are to be normal.

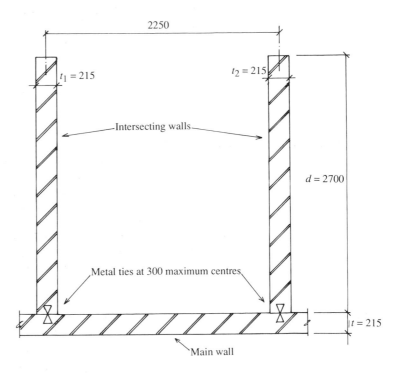

Figure 4.30   *Plan on wall*

The effective height $h_{ef} = 0.75h = 0.75 \times 3450 = 2587.5\,\text{mm}$. The length of the intersecting walls is greater than $10t = 10 \times 215 = 2150\,\text{mm}$ and their thickness is not less than the main wall, $t = 215\,\text{mm}$; therefore they may be considered to provide lateral support in the vertical direction. The degree of support will be simple since the intersecting walls are only tied and not bonded to the main wall. Therefore the effective length is the clear distance between simple lateral supports, that is $l_{ef} = L - t_1 = 2250 - 215 = 2035\,\text{mm}$. As the effective length of 2035 mm is less than the effective height of 2587.5 mm, it will govern the slenderness ratio.

Furthermore, since the length of the intersecting walls is greater than $3t$ they may be considered as equivalent stiffening piers for the purpose of determining the effective thickness:

$$\frac{\text{Equivalent pier spacing}}{\text{Equivalent pier width}} = \frac{2250}{215} = 10.47$$

Equivalent pier thickness $t_p = 3t = 3 \times 215 = 645\,mm$

$$\frac{\text{Equivalent pier thickness}}{\text{Wall thickness}} = \frac{t_p}{t} = \frac{645}{215} = 3$$

Therefore, by interpolation from Table 4.7, the stiffness coefficient K is 1.38. The effective thickness $t_{ef} = tK = 215 \times 1.38 = 296.7\,mm$. Thus the slenderness ratio is given by

$$SR = \frac{\text{effective length}}{\text{effective thickness}} = \frac{l_{ef}}{t_{ef}} = \frac{2035}{296.7} = 6.86 < 27$$

This is satisfactory. From Table 4.7, the capacity reduction factor β is 1.0 without eccentricity.

From Table 4.5a, the masonry characteristic strength $f_k = 8.5\,N/mm^2$. The material partial safety factor $\gamma_m = 3.5$. Thus the ultimate vertical design strength is

$$\frac{\beta t f_k}{\gamma_m} = \frac{1.0 \times 215 \times 8.5}{3.5} = 522.14\,N/mm = 522.14\,kN \text{ per metre run}$$

4.11 Concentrated loads

Concentrated loads can occur at beam, truss or lintel bearings. Whilst these produce relatively high stress concentrations over a small plan area, they are usually rapidly dispersed through the wall construction below. It is accepted that bearing stresses produced by concentrated loads of a purely local nature may safely exceed the allowable design stress for a uniformly distributed load.

Reference should be made to BS 5628 Part 1 for guidance on the three types of bearing condition which permit the normal design stresses to be exceeded by 1.25, 1.5 and 2 times respectively.

4.12 References

BS 187 1978 Specification for calcium silicate (sand lime and flint lime) bricks.

BS 1243 1978 Specification for metal ties for cavity wall construction.

BS 3921 1985 Specification for clay bricks.

BS 4721 1981 (1986) Specification for ready mixed building mortars.

BS 5390 1976 (1984) Code of practice for stone masonry.

BS 5628 Code of practice for use of masonry.
   Part 1 1978 (1985) Structural use of unreinforced masonry.
   Part 3 1985 Materials and components, design and workmanship.

BS 6073 1981 Precast concrete masonry units.
   Part 1 Specification for precast masonry units.
   Part 2 Method for specifying precast concrete masonry units.

Structural Masonry Designers' Manual. W.G. Curtin, G. Shaw, J.K. Beck and W.A. Bray. 2nd edn. BSP Professional Books, 1987.

Structural Masonry Detailing. W.G. Curtin, G. Shaw, J.K. Beck and G.I. Parkinson. BSP Professional Books, 1984.

For further information contact:

Brick Development Association, Woodside House, Winkfield, Windsor, Berks, SL4 2DX.

5 Steel elements

5.1 Structural design of steelwork

At present there are two British Standards devoted to the design of structural steel elements:

BS 449   The use of structural steel in building.

BS 5950   Structural use of steelwork in building.

The former employs permissible stress analysis whilst the latter is based upon limit state philosophy. Since it is intended that BS 5950 will eventually replace BS 449, the designs contained in this manual will be based upon BS 5950.

There are to be nine parts to BS 5950:

Part 1   Code of practice for design in simple and continuous construction: hot rolled sections.

Part 2   Specification for materials, fabrication and erection: hot rolled sections.

Part 3   Code of practice for design in composite construction.

Part 4   Code of practice for design of floors with profiled steel sheeting.

Part 5   Code of practice for design of cold formed sections.

Part 6   Code of practice for design in light gauge sheeting, decking and cladding.

Part 7   Specification for materials and workmanship: cold formed sections.

Part 8   Code of practice for design of fire protection for structural steelwork.

Part 9   Code of practice for stressed skin design.

Calculations for the majority of steel members contained in building and allied structures are usually based upon the guidance given in Part 1 of the standard. This manual will therefore be related to that part.

Requirements for the fabrication and erection of structural steelwork are given in Part 2 of the standard. The designer should also be familiar with these, so that he can take into account any which could influence his design.

For information on all aspects of bridge design, reference should be made to BS 5400, 'Steel, concrete and composite bridges'.

The design of a steel structure may be divided into two stages. First the size of the individual members is determined in relation to the induced forces and bending moments. Then all necessary bolted or welded connections are designed so that they are capable of transmitting the forces and bending moments. In this manual we will concentrate on the design of the main structural elements.

Three methods of design are included in BS 5950 Part 1:

Simple design   This method applies to structures in which the end connections between members are such that they cannot develop any significant restraint moments. Thus, for the purpose of design, the structure may be considered to be pin-jointed on the basis of the following assumptions:

(a) All beams are simply supported.
(b) All connections are designed to resist only resultant reactions at the appropriate eccentricity.
(c) Columns are subjected to loads applied at the appropriate eccentricity.
(d) Resistance to sway, such as that resulting from lateral wind loads, is provided by either bracing, shear walls or core walls.

Rigid design   In this method the structure is considered to be rigidly jointed such that it behaves as a continuous framework. Therefore the connections must be capable of transmitting both forces and bending moments. Portal frames are designed in this manner using either elastic or plastic analysis.

Semi-rigid design   This is an empirical method, seldom adopted, which permits partial interaction between beams and columns to be assumed provided that certain stated parameters are satisfied.

The design of steel elements dealt with in this manual will be based upon the principles of simple design.

It is important to appreciate that an economic steel design is not necessarily that which uses the least weight of steel. The most economical solution will be that which produces the lowest overall cost in terms of materials, detailing, fabrication and erection.

5.2 Symbols

The symbols used in BS 5950 and which are relevant to this manual are as follows:

A area
A_g gross sectional area of steel section
A_v shear area (sections)
B breadth of section
b outstand of flange
b_1 stiff bearing length
D depth of section

d　depth of web

E　modulus of elasticity of steel

e　eccentricity

F_c　ultimate applied axial load

F_v　shear force (sections)

I_x　second moment of area about the major axis

I_y　second moment of area about the minor axis

L　length of span

L_E　effective length

M　larger end moment

M_A　maximum moment on the member or portion of the member under consideration

M_b　buckling resistance moment (lateral torsional)

M_{cx}, M_{cy}　moment capacity of section about the major and minor axes in the absence of axial load

M_e　eccentricity moment

M_o　mid-length moment on a simply supported span equal to the unrestrained length

M_u　ultimate moment

M_x　maximum moment occurring between lateral restraints on a beam

$\bar{M}$　equivalent uniform moment

m　equivalent uniform moment factor

n　slenderness correction factor

P_c　compression resistance of column

P_{crip}　ultimate web bearing capacity

P_v　shear capacity of a section

p_b　bending strength

p_c　compressive strength

P_w　buckling resistance of an unstiffened web

p_y　design strength of steel

r_x, r_y　radius of gyration of a member about its major and minor axes

S_x, S_y　plastic modulus about the major and minor axes

T　thickness of a flange or leg

t　thickness of a web or as otherwise defined in a clause

u　buckling parameter of the section

v　slenderness factor for beam

x　torsional index of section

Z_x, Z_y　elastic modulus about the major and minor axes

β　ratio of smaller to larger end moment

γ_f　overall load factor

γ_ℓ　load variation factor: function of $\gamma_{\ell 1}$ and $\gamma_{\ell 2}$

γ_m　material strength factor

γ　ratio M/M_0, that is the ratio of the larger end moment to the mid-length moment on a simply supported span equal to the unrestrained length

δ　deflection

ε　constant $(275/p_y)^{1/2}$

λ slenderness, that is the effective length divided by the radius of gyration

λ_{LT} equivalent slenderness

5.3 Definitions

The following definitions which are relevant to this manual have been abstracted from BS 5950 Part 1:

Beam   A member predominantly subject to bending.

Buckling resistance   Limit of force or moment which a member can withstand without buckling.

Capacity   Limit of force or moment which may be applied without causing failure due to yielding or rupture.

Column   A vertical member of a structure carrying axial load and possibly moments.

Compact cross-section   A cross-section which can develop the plastic moment capacity of the section but in which local buckling prevents rotation at constant moment.

Dead load   All loads of constant magnitude and position that act permanently, including self-weight

Design strength   The yield strength of the material multiplied by the appropriate partial factor.

Effective length   Length between points of effective restraint of a member multiplied by a factor to take account of the end conditions and loading.

Elastic design   Design which assumes no redistribution of moments due to plastic rotation of a section throughout the structure.

Empirical method   Simplified method of design justified by experience or testing.

Factored load   Specified load multiplied by the relevant partial factor.

H-section   A section with one central web and two equal flanges which has an overall depth not greater than 1.2 times the width of the flange.

I-section   Section with central web and two equal flanges which has an overall depth greater than 1.2 times the width of the flange.

Imposed load   Load on a structure or member other than wind load, produced by the external environment and intended occupancy or use.

Lateral restraint   For a beam: restraint which prevents lateral movement of the compression flange. For a column: restraint which prevents lateral movement of the member in a particular plane.

Plastic cross-section   A cross-section which can develop a plastic hinge with sufficient rotation capacity to allow redistribution of bending moments within the structure.

Plastic design   Design method assuming redistribution of moment in continuous construction.

Semi-compact cross-section   A cross-section in which the stress in the extreme fibres should be limited to yield because local buckling would prevent development of the plastic moment capacity in the section.

Serviceability limit states   Those limit states which when exceeded can lead to the structure being unfit for its intended use.

Slender cross-section   A cross-section in which yield of the extreme fibres cannot be attained because of premature local buckling.

Slenderness   The effective length divided by the radius of gyration.

Strength   Resistance to failure by yielding or buckling.

Strut   A member of a structure carrying predominantly compressive axial load.

Ultimate limit state   That state which if exceeded can cause collapse of part or the whole of the structure.

5.4   Steel grades and sections

As mentioned in Chapter 1, steel sections are produced by rolling the steel, whilst hot, into various standard profiles. The quality of the steel that is used must comply with BS 4360 'Specification for weldable structural steels', which designates four basic grades for steel: 40, 43, 50 and 55. (It should be noted that grade 40 steel is not used for structural purposes.) These basic grades are further classified in relation to their ductility, denoted by suffix letters A, B, C and so on. These in turn give grades 43A, 43B, 43C and so on. The examples in this manual will, for simplicity, be based on the use of grade 43A steel.

It is eventually intended to replace the present designations with grade references related to the yield strength of the steel. Thus, for example, grade 43A steel will become grade 275A since it has a yield stress of 275 N/mm^2.

The dimensions and geometric properties of the various hot rolled sections are obtained from the relevant British Standards. Those for universal beam (UB) sections, universal column (UC) sections, rolled steel joist (RSJ) sections and rolled steel channel (RSC) sections are given in BS 4 Part 1. Structural hollow sections and angles are covered by BS 4848 Part 2 and Part 4 respectively. It is eventually intended that BS 4 Part 1 will also become part of BS 4848.

Cold formed steel sections produced from light gauge plate, sheet or strip are also available. Their use is generally confined to special applications and the production of proprietary roof purlins and sheeting rails. Guidance on design using cold formed sections is given in BS 5950 Part 5.

5.5   Design philosophy

The design approach employed in BS 5950 is based on limit state philosophy. The fundamental principles of the philosophy were explained in Chapter 3 in the context of concrete design. In relation to steel structures, some of the ultimate and serviceability limit states (ULSs and SLSs) that may have to be considered are as follows

Ultimate limit states

Strength   The individual structural elements should be checked to ensure that they will not yield, rupture or buckle under the influence of the ultimate design loads, forces, moments and so on. This will entail checking beams for the ULSs of bending and shear, and columns for a compressive ULS and when applicable a bending ULS.

Stability   The building or structural framework as a whole should be checked to ensure that the applied loads do not induce excessive sway or cause overturning.

Fracture due to fatigue   Fatigue failure could occur in a structure that is repeatedly subjected to rapid reversal of stress. Connections are particularly prone to such failure. In the majority of building structures, changes in stress are gradual. However, where dynamic loading could occur, such as from travelling cranes, the risk of fatigue failure should be considered.

Brittle failure   Sudden failure due to brittle fracture can occur in steel-work exposed to low temperatures; welded structures are particularly susceptible. Since the steel members in most building frames are protected from the weather, they are not exposed to low temperatures and therefore brittle fracture need not be considered. It is more likely to occur in large welded structures, such as bridges, which are exposed to the extremes of winter temperature. In such circumstances, it is necessary to select steel of adequate notch ductility and to devise details that avoid high stress concentrations.

Serviceability limit states

Deflection   Adequate provision must be made to ensure that excessive deflection which could adversely effect any components or finishes supported by the steel members does not occur.

Corrosion and durability   Corrosion induced by atmospheric or chemical conditions can adversely affect the durability of a steel structure. The designer must therefore specify a protective treatment suited to the location of the structure. Guidance on the selection of treatments is given in BS 5493 'Code of practice for protective coating of iron and steel structures against corrosion'. Certain classes of grade 50 steel are also available with weather resistant qualities, indicated by the prefix WR, for example WR 50A. Such steel when used in a normal external environment does not need any additional surface protection. An oxide skin forms on the surface of the steel, preventing further corrosion. Provided that the self-coloured appearance is aesthetically acceptable, consideration may be given to its use in situations where exposed steel is permitted, although it should be borne in mind that it is more expensive than ordinary steel.

Fire protection   Due consideration should also be given to the provision of adequate protection to satisfy fire regulations. Traditionally fire protection was provided by casing the steelwork in concrete. Nowadays a number of lightweight alternatives are available in the form of dry sheet material, plaster applied to metal lathing, or plaster sprayed directly on to the surface of the steel. Intumescent paints are also marketed which froth when heated to produce a protective insulating layer on the surface of the steel.

Since this manual is concerned with the design of individual structural elements, only the strength ULS and the deflection SLS will be considered further.

5.6   Safety factors

In a similar fashion to concrete and masonry design, partial safety factors are once again applied separately to the loads and material stresses. Initially BS 5950 introduces a third factor, γ_p, related to structural performance. The factors given in BS 5950 are as follows:

γ_ℓ for load
γ_p for structural performance
γ_m for material strength.

However, factors γ_ℓ and γ_p when multiplied together give a single partial safety factor for load of γ_f. Hence the three partial safety factors reduce to the usual two of γ_f and γ_m.

5.7   Loads

The basic loads are referred to in BS 5950 as specified loads rather than characteristic loads. They need to be multiplied by the relevant partial safety factor for load γ_f to arrive at the design load.

5.7.1   Specified loads

These are the same as the characteristic loads of dead, imposed and wind previously defined in Chapters 3 and 4 in the context of concrete and masonry design.

5.7.2   Partial safety factors for load

To arrive at the design load, the respective specified loads are multiplied by a partial safety factor γ_f in relation to the limit state being considered:

$$\text{Design load} = \gamma_f \times \text{specified load}$$

5.7.3   Ultimate design load

The partial safety factors for the ULS load combinations are given in Table 2 of BS 5950. For the beam and column examples contained in this manual, only the values for the dead and imposed load combination are required, which are 1.4 and 1.6 respectively. Thus the ultimate design load for the dead plus imposed combination would be as follows:

$$\text{Ultimate design load} = \gamma_f \times \text{dead load} + \gamma_f \times \text{imposed load}$$
$$= 1.4 \times \text{dead load} + 1.6 \times \text{imposed load}$$

5.7.4 Serviceability design load

For the purpose of checking the deflection SLS, the partial safety factor γ_f may be taken as unity. Furthermore, in accordance with BS 5950, the deflection of a beam need only be checked for the effect of imposed loading. Hence the serviceability design load for checking the deflection of a steel beam is simply the specified imposed load. This differs from the design of timber and concrete beams, for which the dead plus imposed load is used to check deflection. However, it is not unreasonable since we are only interested in controlling the deflection of steel beams to avoid damage to finishes, and the dead load deflection will already have taken place before these are applied. If for reasons of appearance it is considered necessary to counteract all or part of the dead load deflection, the beam could be pre-cambered.

5.8   Material properties

The ultimate design strength p_y for the most common types of structural steel are given in BS 5950 Table 6, from which those for grade 43 steel are shown here in Table 5.1. They incorporate the material partial safety factor γ_m in the specified values. Therefore the strength may be obtained directly from the table without further modification. For beam and column sections the material thickness referred to in the table should be taken as the flange thickness.

Table 5.1   Design strength p_y of grade 43 steel

Thickness less than or equal to (mm)	p_y for rolled sections, plates and hollow sections (N/mm^2)
16	275
40	265
63	255
100	245

The modulus of elasticity E, for deflection purposes, may be taken as 205 kN/mm^2 for all grades of steel.

5.9   Section properties

Dimensions and geometric properties for the hot rolled steel sections commonly available for use as beams and columns are tabulated in BS 4 Part 1. Similar tables expanded to include a number of useful design constants are also published by the Steel Construction Institute. These are contained in their *Steelwork Design Guide to BS 5950: Part 1*, Volume 1, *Section Properties, Member Capacities*. Tables 5.2 and 5.3 given here are typical examples from that publication, reproduced by kind permission of the director of the Steel Construction Institute. Complete copies of the guide can be obtained from the Institute at Silwood Park, Ascot, Berkshire, SL5 7QN.

Table 5.2 relates to universal beam (UB) sections, as illustrated in Figure 5.1, and Table 5.3 to universal column (UC) sections, as illustrated in Figure 5.2. The use of these tables in relation to the design of beams and columns will be explained in the appropriate sections of this chapter.

Whilst the UB sections are primarily intended for use as beams, they can if desired be used as columns; this is often the case in portal frame construction. Similarly the UC sections are intended for use as columns but can also be used as beams. However, because they have a stocky cross-section they do not lend themselves as readily to such an alternative use.

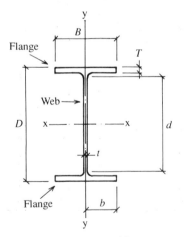

Figure 5.1   *Universal beam cross-section*

Figure 5.2   *Universal column cross-section*

5.10   Beams

The main structural design requirements for which steel beams should be examined as as follows:

(a) Bending ULS

Table 5.2   Universal beams (abstracted from the *Steelwork Design Guide to BS 5950: Part 1*, published by the Steel Construction Institute)

(a) Dimensions

Designation Serial size (mm)	Mass per metre (kg)	Depth of section D (mm)	Width of section B (mm)	Thickness Web t (mm)	Thickness Flange T (mm)	Root radius r (mm)	Depth between fillets d (mm)	Ratios for local buckling Flange b/T	Ratios for local buckling Web d/t	End clearance C (mm)	Notch N (mm)	Notch n (mm)	Surface area Per metre (m²)	Surface area per tonne (m²)
914 × 419	388	920.5	420.5	21.5	36.6	24.1	799.1	5.74	37.2	13	210	62	3.44	8.86
	343	911.4	418.5	19.4	32.0	24.1	799.1	6.54	41.2	12	210	58	3.42	9.96
914 × 305	289	926.6	307.8	19.6	32.0	19.1	824.5	4.81	42.1	12	156	52	3.01	10.4
	253	918.5	305.5	17.3	27.9	19.1	824.5	5.47	47.7	11	156	48	2.99	11.8
	224	910.3	304.1	15.9	23.9	19.1	824.5	6.36	51.9	10	156	44	2.97	13.3
	201	903.0	303.4	15.2	20.2	19.1	824.5	7.51	54.2	10	156	40	2.96	14.7
838 × 292	226	850.9	293.8	16.1	26.8	17.8	761.7	5.48	47.3	10	150	46	2.81	12.5
	194	840.7	292.4	14.7	21.7	17.8	761.7	6.74	51.8	9	150	40	2.79	14.4
	176	834.9	291.6	14.0	18.8	17.8	761.7	7.76	54.5	9	150	38	2.78	15.8
762 × 267	197	769.6	268.0	15.6	25.4	16.5	685.8	5.28	44.0	10	138	42	2.55	13.0
	173	762.0	266.7	14.3	21.6	16.5	685.8	6.17	48.0	9	138	40	2.53	14.6
	147	753.9	265.3	12.9	17.5	16.5	685.8	7.58	53.2	8	138	36	2.51	17.1
686 × 254	170	692.9	255.8	14.5	23.7	15.2	615.1	5.40	42.4	9	132	40	2.35	13.8
	152	687.6	254.5	13.2	21.0	15.2	615.1	6.06	46.6	9	132	38	2.34	15.4
	140	683.5	253.7	12.4	19.0	15.2	615.1	6.68	49.6	8	132	36	2.33	16.6
	125	677.9	253.0	11.7	16.2	15.2	615.1	7.81	52.6	8	132	32	2.32	18.5
610 × 305	238	633.0	311.5	18.6	31.4	16.5	537.2	4.96	28.9	11	158	48	2.45	10.3
	179	617.5	307.0	14.1	23.6	16.5	537.2	6.50	38.1	9	158	42	2.41	13.4
	149	609.6	304.8	11.9	19.7	16.5	537.2	7.74	45.1	8	158	38	2.39	16.0
610 × 229	140	617.0	230.1	13.1	22.1	12.7	547.3	5.21	41.8	9	120	36	2.11	15.0
	125	611.9	229.0	11.9	19.6	12.7	547.3	5.84	46.0	8	120	34	2.09	16.8
	113	607.3	228.2	11.2	17.3	12.7	547.3	6.60	48.9	8	120	32	2.08	18.4
	101	602.2	227.6	10.6	14.8	12.7	547.3	7.69	51.6	7	120	28	2.07	20.5
533 × 210	122	544.6	211.9	12.8	21.3	12.7	476.5	4.97	37.2	8	110	36	1.89	15.5
	109	539.5	210.7	11.6	18.8	12.7	476.5	5.60	41.1	8	110	32	1.88	17.2
	101	536.7	210.1	10.9	17.4	12.7	476.5	6.04	43.7	7	110	32	1.87	18.5
	92	533.1	209.3	10.2	15.6	12.7	476.5	6.71	46.7	7	110	30	1.86	20.2
	82	528.3	208.7	9.6	13.2	12.7	476.5	7.91	49.6	7	110	26	1.85	22.6
457 × 191	98	467.4	192.8	11.4	19.6	10.2	407.9	4.92	35.8	8	102	30	1.67	17.0
	89	463.6	192.0	10.6	17.7	10.2	407.9	5.42	38.5	7	102	28	1.66	18.6
	82	460.2	191.3	9.9	16.0	10.2	407.9	5.98	41.2	7	102	28	1.65	20.1
	74	457.2	190.5	9.1	14.5	10.2	407.9	6.57	44.8	7	102	26	1.64	22.2
	67	453.6	189.9	8.5	12.7	10.2	407.9	7.48	48.0	6	102	24	1.63	24.4
457 × 152	82	465.1	153.5	10.7	18.9	10.2	407.0	4.06	38.0	7	82	30	1.51	18.4
	74	461.3	152.7	9.9	17.0	10.2	407.0	4.49	41.1	7	82	28	1.50	20.2
	67	457.3	151.9	9.1	15.0	10.2	407.0	5.06	44.7	7	82	26	1.49	22.2
	60	454.7	152.9	8.0	13.3	10.2	407.0	5.75	51.0	6	84	24	1.49	24.8
	52	449.8	152.4	7.6	10.9	10.2	407.0	6.99	53.6	6	84	22	1.48	28.4

USE BOOK TABLES

Table 5.2   Universal beams *continued* (abstracted from the *Steelwork Design Guide to BS 5950: Part 1*, published by the Steel Construction Institute)

(b) Properties

Designation		Second moment of area		Radius of gyration		Elastic modulus		Plastic modulus		Buckling parameter	Torsional index	Warping constant	Torsional constant	Area of section
Serial size	Mass per metre	Axis $x-x$	Axis $y-y$	Axis $x-x$	Axis $y-y$	Axis $x-x$	Axis $y-y$	Axis $x-x$	Axis $y-y$	u	x	H	J	A
(mm)	(kg)	(cm^4)	(cm^4)	(cm)	(cm)	(cm^3)	(cm^3)	(cm^3)	(cm^3)			(dm^6)	(cm^4)	(cm^2)
914 × 419	388	719 000	45 400	38.1	9.58	15 600	2160	17 700	3340	0.884	26.7	88.7	1730	494
	343	625 000	39 200	37.8	9.46	13 700	1870	15 500	2890	0.883	30.1	75.7	1190	437
914 × 305	289	505 000	15 600	37.0	6.51	10 900	1010	12 600	1600	0.867	31.9	31.2	929	369
	253	437 000	13 300	36.8	6.42	9 510	872	10 900	1370	0.866	36.2	26.4	627	323
	224	376 000	11 200	36.3	6.27	8 260	738	9 520	1160	0.861	41.3	22.0	421	285
	201	326 000	9 430	35.6	6.06	7 210	621	8 360	983	0.853	46.8	18.4	293	256
838 × 292	226	340 000	11 400	34.3	6.27	7 990	773	9 160	1210	0.87	35.0	19.3	514	289
	194	279 000	9 070	33.6	6.06	6 650	620	7 650	974	0.862	41.6	15.2	307	247
	176	246 000	7 790	33.1	5.90	5 890	534	6 810	842	0.856	46.5	13.0	222	224
762 × 267	197	240 000	8 170	30.9	5.71	6 230	610	7 170	959	0.869	33.2	11.3	405	251
	173	205 000	6 850	30.5	5.57	5 390	513	6 200	807	0.864	38.1	9.38	267	220
	147	169 000	5 470	30.0	5.39	4 480	412	5 170	649	0.857	45.1	7.41	161	188
686 × 254	170	170 000	6 620	28.0	5.53	4 910	518	5 620	810	0.872	31.8	7.41	307	217
	152	150 000	5 780	27.8	5.46	4 370	454	5 000	710	0.871	35.5	6.42	219	194
	140	136 000	5 180	27.6	5.38	3 990	408	4 560	638	0.868	38.7	5.72	169	179
	125	118 000	4 380	27.2	5.24	3 480	346	4 000	542	0.862	43.9	4.79	116	160
610 × 305	238	208 000	15 800	26.1	7.22	6 560	1020	7 460	1570	0.886	21.1	14.3	788	304
	179	152 000	11 400	25.8	7.08	4 910	743	5 520	1140	0.886	27.5	10.1	341	228
	149	125 000	9 300	25.6	6.99	4 090	610	4 570	937	0.886	32.5	8.09	200	190
610 × 229	140	112 000	4 510	25.0	5.03	3 630	392	4 150	612	0.875	30.5	3.99	217	178
	125	98 600	3 930	24.9	4.96	3 220	344	3 680	536	0.873	34.0	3.45	155	160
	113	87 400	3 440	24.6	4.88	2 880	301	3 290	470	0.87	37.9	2.99	112	144
	101	75 700	2 910	24.2	4.75	2 510	256	2 880	400	0.863	43.0	2.51	77.2	129
533 × 210	122	76 200	3 390	22.1	4.67	2 800	320	3 200	501	0.876	27.6	2.32	180	156
	109	66 700	2 940	21.9	4.60	2 470	279	2 820	435	0.875	30.9	1.99	126	139
	101	61 700	2 690	21.8	4.56	2 300	257	2 620	400	0.874	33.1	1.82	102	129
	92	55 400	2 390	21.7	4.51	2 080	229	2 370	356	0.872	36.4	1.60	76.2	118
	82	47 500	2 010	21.3	4.38	1 800	192	2 060	300	0.865	41.6	1.33	51.3	104
457 × 191	98	45 700	2 340	19.1	4.33	1 960	243	2 230	378	0.88	25.8	1.17	121	125
	89	41 000	2 090	19.0	4.28	1 770	217	2 010	338	0.879	28.3	1.04	90.5	114
	82	37 100	1 870	18.8	4.23	1 610	196	1 830	304	0.877	30.9	0.923	69.2	105
	74	33 400	1 670	18.7	4.19	1 460	175	1 660	272	0.876	33.9	0.819	52.0	95.0
	67	29 400	1 450	18.5	4.12	1 300	153	1 470	237	0.873	37.9	0.706	37.1	85.4
457 × 152	82	36 200	1 140	18.6	3.31	1 560	149	1 800	235	0.872	27.3	0.569	89.3	104
	74	32 400	1 010	18.5	3.26	1 410	133	1 620	209	0.87	30.0	0.499	66.6	95.0
	67	28 600	878	18.3	3.21	1 250	116	1 440	182	0.867	33.6	0.429	47.5	85.4
	60	25 500	794	18.3	3.23	1 120	104	1 280	163	0.869	37.5	0.387	33.6	75.9
	52	21 300	645	17.9	3.11	949	84.6	1 090	133	0.859	43.9	0.311	21.3	66.5

Table 5.3 Universal columns (abstracted from the *Steelwork Design Guide to BS 5950: Part 1*, published by the Steel Construction Institute)

(a) Dimensions

Designation		Depth of section D	Width of section B	Thickness		Root radius	Depth between fillets	Ratios for local buckling		Dimensions for detailing			Surface area	
Serial size	Mass per metre			Web t	Flange T	r	d	Flange b/T	Web d/t	End clearance C	Notch N	n	Per metre	per tonne
(mm)	(kg)	(mm)	(mm)	(mm)	(mm)	(mm)	(mm)			(mm)	(mm)	(mm)	(m^2)	(m^2)
356 × 406	634	474.7	424.1	47.6	77.0	15.2	290.2	2.75	6.10	26	200	94	2.52	3.98
	551	455.7	418.5	42.0	67.5	15.2	290.2	3.10	6.91	23	200	84	2.48	4.49
	467	436.6	412.4	35.9	58.0	15.2	290.2	3.56	8.08	20	200	74	2.42	5.19
	393	419.1	407.0	30.6	49.2	15.2	290.2	4.14	9.48	17	200	66	2.38	6.05
	340	406.4	403.0	26.5	42.9	15.2	290.2	4.70	11.0	15	200	60	2.35	6.90
	287	393.7	399.0	22.6	36.5	15.2	290.2	5.47	12.8	13	200	52	2.31	8.06
	235	381.0	395.0	18.5	30.2	15.2	290.2	6.54	15.7	11	200	46	2.28	9.70
COLCORE	477	427.0	424.4	48.0	53.2	15.2	290.2	3.99	6.05	26	200	70	2.43	5.09
356 × 368	202	374.7	374.4	16.8	27.0	15.2	290.2	6.93	17.3	10	190	44	2.19	10.8
	177	368.3	372.1	14.5	23.8	15.2	290.2	7.82	20.0	9	190	40	2.17	12.3
	153	362.0	370.2	12.6	20.7	15.2	290.2	8.94	23.0	8	190	36	2.16	14.1
	129	355.6	368.3	10.7	17.5	15.2	290.2	10.5	27.1	7	190	34	2.14	16.6
305 × 305	283	365.3	321.8	26.9	44.1	15.2	246.6	3.65	9.17	15	158	60	1.94	6.85
	240	352.6	317.9	23.0	37.7	15.2	246.6	4.22	10.7	14	158	54	1.90	7.93
	198	339.9	314.1	19.2	31.4	15.2	246.6	5.00	12.8	12	158	48	1.87	9.43
	158	327.2	310.6	15.7	25.0	15.2	246.6	6.21	15.7	10	158	42	1.84	11.6
	137	320.5	308.7	13.8	21.7	15.2	246.6	7.11	17.9	9	158	38	1.82	13.3
	118	314.5	306.8	11.9	18.7	15.2	246.6	8.20	20.7	8	158	34	1.81	5.3
	97	307.8	304.8	9.9	15.4	15.2	246.6	9.90	24.9	7	158	32	1.79	18.4
254 × 254	167	289.1	264.5	19.2	31.7	12.7	200.3	4.17	10.4	12	134	46	1.58	9.44
	132	276.4	261.0	15.6	25.3	12.7	200.3	5.16	12.8	10	134	40	1.54	11.7
	107	266.7	258.3	13.0	20.5	12.7	200.3	6.30	15.4	9	134	34	1.52	14.2
	89	260.4	255.9	10.5	17.3	12.7	200.3	7.40	19.1	7	134	32	1.50	16.9
	73	254.0	254.0	8.6	14.2	12.7	200.3	8.94	23.3	6	134	28	1.49	20.3
203 × 203	86	222.3	208.8	13.0	20.5	10.2	160.9	5.09	12.4	9	108	32	1.24	14.4
	71	215.9	206.2	10.3	17.3	10.2	160.9	5.96	15.6	7	108	28	1.22	17.2
	60	209.6	205.2	9.3	14.2	10.2	160.9	7.23	17.3	7	108	26	1.20	20.1
	52	206.2	203.9	8.0	12.5	10.2	160.9	8.16	20.1	6	108	24	1.19	23.0
	46	203.2	203.2	7.3	11.0	10.2	160.9	9.24	22.0	6	108	22	1.19	25.8
152 × 152	37	161.8	154.4	8.1	11.5	7.6	123.5	6.71	15.2	6	84	20	0.912	24.6
	30	157.5	152.9	6.6	9.4	7.6	123.5	8.13	18.7	5	84	18	0.9	30.0
	23	152.4	152.4	6.1	6.8	7.6	123.5	11.2	20.2	5	84	16	0.889	38.7

Table 5.3 Universal columns *continued* (abstracted from the *Steelwork Design Guide to BS 5950: Part 1*, published by the Steel Construction Institute)

(b) Properties

Designation		Second moment of area		Radius of gyration		Elastic modulus		Plastic modulus		Buckling parameter	Torsional index	Warping constant	Torsional constant	Area of section
Serial size	Mass per metre	Axis $x-x$	Axis $y-y$	Axis $x-x$	Axis $y-y$	Axis $x-x$	Axis $y-y$	Axis $x-x$	Axis $y-y$	u	x	H	J	A
(mm)	(kg)	(cm^4)	(cm^4)	(cm)	(cm)	(cm^3)	(cm^3)	(cm^3)	(cm^3)			(dm^6)	(cm^4)	(cm^2)
356 × 406	634	275 000	98 200	18.5	11.0	11 600	4630	14 200	7110	0.843	5.46	38.8	13 700	808
	551	227 000	82 700	18.0	10.9	9 960	3950	12 100	6060	0.841	6.05	31.1	9 240	702
	467	183 000	67 900	17.5	10.7	8 390	3290	10 000	5040	0.839	6.86	24.3	5 820	595
	393	147 000	55 400	17.1	10.5	7 000	2720	8 230	4160	0.837	7.86	19.0	3 550	501
	340	122 000	46 800	16.8	10.4	6 030	2320	6 990	3540	0.836	8.85	15.5	2 340	433
	287	100 000	38 700	16.5	10.3	5 080	1940	5 820	2950	0.835	10.2	12.3	1 440	366
	235	79 100	31 000	16.2	10.2	4 150	1570	4 690	2380	0.834	12.1	9.54	812	300
COLCORE	477	172 000	68 100	16.8	10.6	8 080	3210	9 700	4980	0.815	6.91	23.8	5 700	607
356 × 368	202	66 300	23 600	16.0	9.57	3 540	1260	3 980	1920	0.844	13.3	7.14	560	258
	177	57 200	20 500	15.9	9.52	3 100	1100	3 460	1670	0.844	15.0	6.07	383	226
	153	48 500	17 500	15.8	9.46	2 680	944	2 960	1430	0.844	17.0	5.09	251	195
	129	40 200	14 600	15.6	9.39	2 260	790	2 480	1200	0.843	19.9	4.16	153	165
305 × 305	283	78 800	24 500	14.8	8.25	4 310	1530	5 100	2340	0.855	7.65	6.33	2 030	360
	240	64 200	20 200	14.5	8.14	3 640	1270	4 250	1950	0.854	8.73	5.01	1 270	306
	198	50 800	16 200	14.2	8.02	2 990	1030	3 440	1580	0.854	10.2	3.86	734	252
	158	38 700	12 500	13.9	7.89	2 370	806	2 680	1230	0.852	12.5	2.86	379	201
	137	32 800	10 700	13.7	7.82	2 050	691	2 300	1050	0.851	14.1	2.38	250	175
	118	27 600	9 010	13.6	7.75	1 760	587	1 950	892	0.851	16.2	1.97	160	150
	97	22 200	7 270	13.4	7.68	1 440	477	1 590	723	0.850	19.3	1.55	91.1	123
254 × 254	167	29 900	9 800	11.9	6.79	2 070	741	2 420	1130	0.852	8.49	1.62	625	212
	132	22 600	7 520	11.6	6.67	1 630	576	1 870	879	0.850	10.3	1.18	322	169
	107	17 500	5 900	11.3	6.57	1 310	457	1 490	695	0.848	12.4	0.894	173	137
	89	14 300	4 850	11.2	6.52	1 100	379	1 230	575	0.849	14.4	0.716	104	114
	73	11 400	3 870	11.1	6.46	894	305	989	462	0.849	17.3	0.557	57.3	92.9
203 × 203	86	9 460	3 120	9.27	5.32	851	299	979	456	0.85	10.2	0.317	138	110
	71	7 650	2 540	9.16	5.28	708	246	802	374	0.852	11.9	0.25	81.5	91.1
	60	6 090	2 040	8.96	5.19	581	199	652	303	0.847	14.1	0.195	46.6	75.8
	52	5 260	1 770	8.90	5.16	510	174	568	264	0.848	15.8	0.166	32.0	66.4
	46	4 560	1 540	8.81	5.11	449	151	497	230	0.846	17.7	0.142	22.2	58.8
152 × 152	37	2 220	709	6.84	3.87	274	91.8	310	140	0.848	13.3	0.04	19.5	47.4
	30	1 740	558	6.75	3.82	221	73.1	247	111	0.848	16.0	0.0306	10.5	38.2
	23	1 260	403	6.51	3.68	166	52.9	184	80.9	0.837	20.4	0.0214	4.87	29.8

(b) Shear ULS

(c) Deflection SLS.

Two other ultimate limit state factors that should be given consideration are:

(d) Web buckling resistance

(e) Web bearing resistance.

However, these are not usually critical under normal loading conditions, and in any case may be catered for by the inclusion of suitably designed web stiffeners.

Let us consider how each of these requirements influences the design of beams.

5.10.1 Bending ULS

When a simply supported beam bends, the extreme fibres above the neutral axis are placed in compression. If the beam is a steel beam this means that the top flange of the section is in compression and correspondingly the bottom flange is in tension. Owing to the combined effect of the resultant compressive loading and the vertical loading, the top flange could tend to deform sideways and twist about its longitudinal axis as illustrated in Figure 5.3. This is termed lateral torsional buckling, and could lead to premature failure of the beam before it reaches its vertical moment capacity.

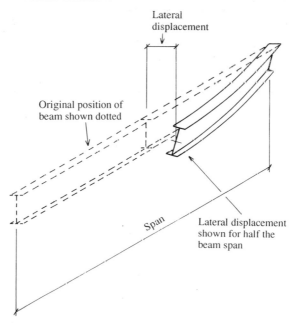

Figure 5.3 *Lateral torsional buckling*

Lateral torsional buckling can be avoided by fully restraining the compression flange along its entire length (Figure 5.4). Alternatively, transverse restraint members can be introduced along the span of the beam (Figure 5.5). These must be at sufficient intervals to prevent lateral torsional buckling occurring between the points of restraint. If neither of these measures are adopted then the beam must be considered as laterally unrestrained and its resistance to lateral torsional buckling should be checked. The requirements that must be fulfilled by both lateral and torsional restraints are described in BS 5950.

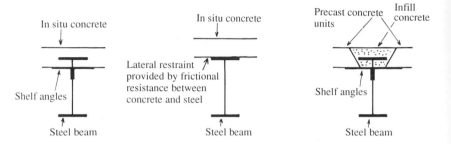

Figure 5.4   *Cross-sections through fully laterally restrained beams*

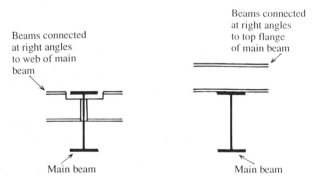

Figure 5.5   *Cross-sections through beams laterally restrained at intervals along their length*

It can be seen from the foregoing that it is necessary to investigate the bending ULS of steel beams in one of two ways: laterally restrained and laterally unrestrained. These are now discussed in turn.

5.10.2   Bending ULS of laterally restrained beams

It has already been shown in Chapter 1 that, in relation to the theory of bending, the elastic moment of resistance (MR) of a steel beam is given by

$$MR = fZ$$

where f is the permissible bending stress value for the steel and Z is the elastic modulus of the section. This assumes that the elastic stress distribution over the depth of the section will be a maximum at the extreme fibres and zero at the neutral axis (NA), as shown in Figure 5.6.

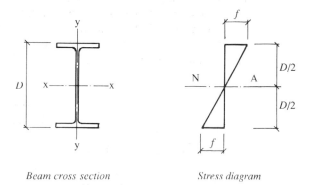

Beam cross section   *Stress diagram*

Figure 5.6   *Elastic stress distribution*

To ensure the adequacy of a particular steel beam, its internal moment of resistance must be equal to or greater than the applied bending moment:

$$MR \geqslant BM$$

This was the method employed in previous Codes of Practice for steel design based upon permissible stress analysis.

In limit state design, advantage is taken of the ability of many steel sections to carry greater loads up to a limit where the section is stressed to yield throughout its depth, as shown in Figure 5.7. The section in such a case is said to have become fully plastic. The moment capacity of such a beam about its major x–x axis would be given by

$$M_{cx} = p_y S_x$$

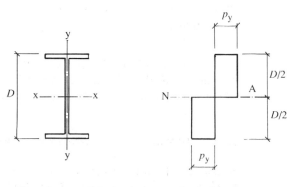

Beam cross-section   *Stress diagram*

Figure 5.7   *Plastic stress distribution*

where p_y is the design strength of the steel, given in Table 5.1, and S_x is the plastic modulus of the section about the major axis, obtained from section tables. In order that plasticity at working load does not occur before the ultimate load is reached, BS 5950 places a limit on the moment capacity of $1.2\,p_y Z$. Thus

$$M_{cx} = p_y S_x \leqslant 1.2\,p_y Z_x$$

The suitability of a particular steel beam would be checked by ensuring that the moment capacity of the section is equal to or greater than the applied ultimate moment M_u:

$$M_{cx} \geqslant M_u$$

The web and flanges of steel sections are comparatively slender in relation to their depth and breadth. Consequently the compressive force induced in a beam by bending could cause local buckling of the web or flange before the full plastic stress is developed. This must not be confused with the previously mentioned lateral torsional buckling, which is a different mode of failure and will be dealt with in the next section. Nor should it be confused with the web buckling ULS discussed in Section 5.10.6.

Local buckling may be avoided by reducing the stress capacity of the section, and hence its moment capacity, relative to its susceptibility to local buckling failure. In this respect steel sections are classified by BS 5950 in relation to the b/T of the flange and the d/t of the web, where b, d, T and t are as previously indicated in Figures 5.1 and 5.2. There are four classes of section:

Class 1   Plastic
Class 2   Compact
Class 3   Semi-compact
Class 4   Slender.

The limiting width to thickness ratios for classes 1, 2 and 3 are given in BS 5950 Table 7, for both beams and columns. Those for rolled beams are listed here in Table 5.4.

Table 5.4   Beam cross-section classification

Limiting proportions	Class of section		
	Plastic	Compact	Semi-compact
b/T	8.5ε	9.5ε	15ε
d/t	79ε	98ε	120ε

The constant $\varepsilon = (275/p_y)^{1/2}$. Hence for grade 43 steel:
When $T \leqslant 16\,\text{mm}$, $\varepsilon = (275/275)^{1/2} = 1$
When $T > 16\,\text{mm}$, $\varepsilon = (275/265)^{1/2} = 1.02$

Slender sections are those whose thickness ratios exceed the limits for semi-compact sections. Their design strength p_y has to be reduced using a stress reduction factor for slender elements, obtained from Table 8 of BS 5950.

The stress distribution and moment capacity for each class of section is shown in Figure 5.8.

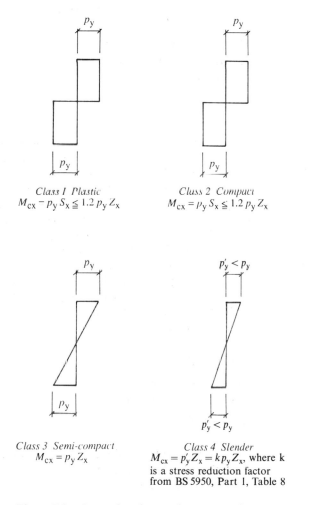

Class 1 Plastic
$M_{cx} = p_y S_x \leq 1.2 \, p_y Z_x$

Class 2 Compact
$M_{cx} = p_y S_x \leq 1.2 \, p_y Z_x$

Class 3 Semi-compact
$M_{cx} = p_y Z_x$

Class 4 Slender
$M_{cx} = p'_y Z_x = k \, p_y Z_x$, where k is a stress reduction factor from BS 5950, Part 1, Table 8

Figure 5.8 *Stress distribution diagrams and moment capacities for section classes*

The examples contained in this manual are based upon the use of grade 43 steel sections. All the UB sections formed from grade 43 steel satisfy either the plastic or the compact classification parameters, and hence the stress reduction factor for slender elements does not apply. Furthermore, their plastic modulus S_x never exceeds 1.2 times their elastic modulus Z_x. Therefore the moment capacity of grade 43 beams will be given by the expression

$$M_{cx} = p_y S_x$$

By rearranging this expression, the plastic modulus needed for a grade 43 UB section to resist a particular ultimate moment may be determined:

$$S_x \text{ required} = \frac{M_u}{p_y}$$

Example 5.1

Steel floor beams arranged as shown in Figure 5.9 support a 150 mm thick reinforced concrete slab which fully restrains the beams laterally. If the floor has to support a specified imposed load of 5 kN/m^2 and reinforced concrete weighs 2400 kg/m^3, determine the size of grade 43 UBs required.

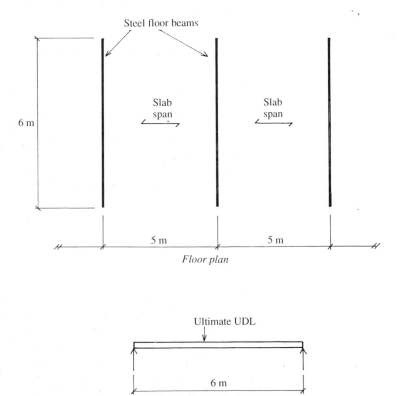

Figure 5.9   *Floor beam arrangement*

Before proceeding to the design of the actual beams it is first necessary to calculate the ultimate design load on an individual beam. This basically follows the procedure explained in Chapter 1, except that partial safety factors for load γ_f need to be applied since we are using limit state design.

Specified dead load 150 mm slab $= 0.15 \times 2400/100 = 3.6 \text{ kN/m}^2$

Specified dead load UDL $= (3.6 \times 6 \times 5) + \text{SW} = 108 + \text{ say } 4 = 112 \text{ kN}$

Specified imposed load $= 5 \, \text{kN/m}^2$

Specified imposed load UDL $= 5 \times 6 \times 5 = 150 \, \text{kN}$

Total ULS design load $= \gamma_f \times$ specified dead load $+ \gamma_f \times$ specified imposed load
$$= 1.4 \times 112 + 1.6 \times 150 = 156.8 + 240 = 396.8 \, \text{kN}$$

Ultimate bending moment $M_u = \dfrac{WL}{8} = \dfrac{396.8 \times 6}{8}$
$$= 297.6 \, \text{kN m} = 297.6 \times 10^6 \, \text{N mm}$$

The ultimate design strength p_y for grade 43 steel sections, from Table 5.1, is 275 N/mm^2 provided that the flange thickness does not exceed 16 mm. If the flange thickness was greater than 16 mm, p_y would reduce to 265 N/mm^2. Hence the plastic modulus is

$$S_x \text{ required} = \frac{M_u}{p_y} = \frac{297.6 \times 10^6}{275} = 1\,082\,182 \, \text{mm}^3 = 1082 \, \text{cm}^3$$

It should be appreciated that the plastic modulus property is always tabulated in cm^3 units.

By reference to Table 5.2, the lightest UB section with a plastic modulus greater than that required is a $457 \times 152 \times 60 \, \text{kg/m}$ UB with an S_x of $1280 \, \text{cm}^3$. It should be noted that the flange thickness of the selected section is 13.3 mm; this is less than 16 mm, and it was therefore correct to adopt a p_y of 275 N/mm^2 in the design. It should also be noted that the self-weight of the section is less than that assumed and therefore no adjustment to the design is necessary; that is,

$$SW = \frac{60}{100} \times 6 = 3.6 \, \text{kN} < 4 \, \text{kN assumed}$$

This section would be adopted provided that it could also satisfy the shear and deflection requirements which will be discussed later.

The design approach employed in Example 5.1 only applies to beams which are fully restrained laterally and are subject to low shear loads. When plastic and compact beam sections are subject to high shear loads their moment capacity reduces because of the interaction between shear and bending. Modified expressions are given in BS 5950 for the moment capacity of beams in such circumstances. However, except for heavily loaded short span beams, this is not usually a problem and it will therefore not be given any further consideration here.

5.10.3 Bending ULS of laterally unrestrained beams

Laterally unrestrained beams are susceptible to lateral torsional buckling failure, and must therefore be designed for a lower moment capacity known as the buckling resistance moment M_b. It is perhaps worth reiterating that torsional buckling is not the same as local buckling, which also needs to be taken into account by reference to the section classification of plastic, compact, semi-compact or slender.

For rolled universal sections or joists BS 5950 offers two alternative approaches – rigorous or conservative – for the assessment of a member's lateral torsional buckling resistance. The rigorous approach may be applied to any form of section acting as a beam, whereas the conservative approach applies only to UB, UC and RSJ sections. Let us therefore consider the implications of each of these approaches with respect to the design of rolled universal sections.

Laterally unrestrained beams, rigorous approach

Unlike laterally restrained beams, it is the section's buckling resistance moment M_b that is usually the criterion rather than its moment capacity M_c. This is given by the following expression:

$$M_b = p_b S_x$$

where p_b is the bending strength and S_x is the plastic modulus of the section about the major axis, obtained from section tables.

The bending strength of laterally unrestrained rolled sections is obtained from BS 5950 Table 11, reproduced here as Table 5.5. It depends on the steel design strength p_y and the equivalent slenderness λ_{LT}, which is derived from the following expression:

$$\lambda_{LT} = nuv\lambda$$

where

- n slenderness correction factor from BS 5950
- u buckling parameter of the section, found from section tables or conservatively taken as 0.9
- v slenderness factor from BS 5950
- λ minor axis slenderness: $\lambda = L_E/r_y$
- L_E effective unrestrained length of the beam
- r_y radius of gyration of the section about its minor axis, from section tables

The effective length L_E should be obtained in accordance with one of the following conditions:

Condition (a). For beams with lateral restraints at the ends only, the value of L_E should be obtained from BS 5950 Table 9, reproduced here as Table 5.6, taking L as the span of the beam. Where the restraint conditions at each end of the beam differ, the mean value of L_E should be taken.

Condition (b). For beams with effective lateral restraints at intervals along their length, the value of L_E should be taken as $1.0L$ for normal loading conditions or $1.2L$ for destabilizing conditions, taking L as the distance between restraints.

Condition (c). For the portion of a beam between one end and the first intermediate restraint, account should be taken of the restraint conditions

Table 5.5 Bending strength p_b (N/mm²) for rolled sections (BS 5950 Part 1 1990 Table 11)

λ_{LT}	p_y								
	245	265	275	325	340	355	415	430	450
30	245	265	275	325	340	355	408	421	438
35	245	265	273	316	328	341	390	402	418
40	238	254	262	302	313	325	371	382	397
45	227	242	250	287	298	309	350	361	374
50	217	231	238	272	282	292	329	338	350
55	206	219	226	257	266	274	307	315	325
60	195	207	213	241	249	257	285	292	300
65	185	196	201	225	232	239	263	269	276
70	174	184	188	210	216	222	242	247	253
75	164	172	176	195	200	205	223	226	231
80	154	161	165	181	186	190	204	208	212
85	144	151	154	168	172	175	188	190	194
90	135	141	144	156	159	162	173	175	178
95	126	131	134	144	147	150	159	161	163
100	118	123	125	134	137	139	147	148	150
105	111	115	117	125	127	129	136	137	139
110	104	107	109	116	118	120	126	127	128
115	97	101	102	108	110	111	117	118	119
120	91	94	96	101	103	104	108	109	111
125	86	89	90	95	96	97	101	102	103
130	81	83	84	89	90	91	94	95	96
135	76	78	79	83	84	85	88	89	90
140	72	74	75	78	79	80	83	84	84
145	68	70	71	74	75	75	78	79	79
150	64	66	67	70	70	71	73	74	75
155	61	62	63	66	66	67	69	70	70
160	58	59	60	62	63	63	65	66	66
165	55	56	57	59	60	60	62	62	63
170	52	53	54	56	56	57	59	59	59
175	50	51	51	53	54	54	56	56	56
180	47	48	49	51	51	51	53	53	53
185	45	46	46	48	49	49	50	50	51
190	43	44	44	46	46	47	48	48	48
195	41	42	42	44	44	44	46	46	46
200	39	40	40	42	42	42	43	44	44
210	36	37	37	38	39	39	40	40	40
220	33	34	34	35	35	36	36	37	37
230	31	31	31	32	33	33	33	34	34
240	29	29	29	30	30	30	31	31	31
250	27	27	27	28	28	28	29	29	29

Table 5.6   Effective length L_E for beams (BS 5950 Part 1 1990 Table 9)

Conditions of restraint at supports		Loading conditions	
		Normal	Destabilizing
Compression flange laterally restrained Beam fully restrained against torsion	Both flanges fully restrained against rotation on plan	$0.7\,L$	$0.85\,L$
	Both flanges partially restrained against rotation on plan	$0.85\,L$	$1.0\,L$
	Both flanges free to rotate on plan	$1.0\,L$	$1.2\,L$
Compression flange laterally unrestrained Both flanges free to rotate on plan	Restraint against torsion provided only by positive connection of bottom flange to supports	$1.0\,L + 2\,D$	$1.2\,L + 2\,D$
	Restraint against torsion provided only by dead bearing of bottom flange on supports	$1.2\,L + 2\,D$	$1.4\,L + 2\,D$

D is the depth of the beam.
L is the span of the beam.

Point load applied by column

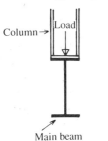

Main beam

(a) Destabilizing detail

Point load applied by column

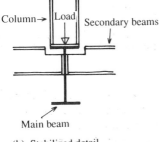

Main beam

(b) Stabilized detail

Figure 5.10   *Destabilizing load*

at the support. Therefore the effective length L_E should be taken as the mean of the value given by condition (b) and the value from Table 5.6 relating to the manner of restraint at the support. In both cases, L is taken as the distance between the restrain and the support.

The destabilizing load referred to in the table exists when the member applying the load to the compression flange can move laterally with the beam in question, as illustrated in Figure 5.10a. This may be avoided by the introduction of stabilizing members such as the secondary beams shown in Figure 5.10b.

The slenderness factor v is obtained from BS 5950 Table 14, reproduced here as Table 5.7, using N and λ/x, where λ is the slenderness, x is the torsional index of the section from section tables, and N is 0.5 for beams with equal flanges.

To check the adequacy of a particular steel beam section, the buckling moment M_b should be compared with the equivalent uniform moment $\bar{M}$:

$$\bar{M} \leqslant M_b$$

where $\bar{M} = mM_A$, m is the equivalent uniform moment factor from BS 5950, and M_A is the maximum moment on the member or portion of the member under consideration.

Table 5.7   Slenderness factor v for flanged beams of uniform section (BS 5950 Part 1 1990 Table 14)

λ/x \\ N	1.0	0.9	0.8	0.7	0.6	0.5	0.4	0.3	0.2	0.1	0.0
0.5	0.79	0.81	0.84	0.88	0.93	1.00	1.11	1.28	1.57	2.20	12.67
1.0	0.78	0.80	0.83	0.87	0.92	0.99	1.10	1.27	1.53	2.11	6.36
1.5	0.77	0.80	0.82	0.86	0.91	0.97	1.08	1.24	1.48	1.98	4.27
2.0	0.76	0.78	0.81	0.85	0.89	0.96	1.06	1.20	1.42	1.84	3.24
2.5	0.75	0.77	0.80	0.83	0.88	0.93	1.03	1.16	1.35	1.70	2.62
3.0	0.74	0.76	0.78	0.82	0.86	0.91	1.00	1.12	1.29	1.57	2.21
3.5	0.72	0.74	0.77	0.80	0.84	0.89	0.97	1.07	1.22	1.46	1.93
4.0	0.71	0.73	0.75	0.78	0.82	0.86	0.94	1.03	1.16	1.36	1.71
4.5	0.69	0.71	0.73	0.76	0.80	0.84	0.91	0.99	1.11	1.27	1.55
5.0	0.68	0.70	0.72	0.75	0.78	0.82	0.88	0.95	1.05	1.20	1.41
5.5	0.66	0.68	0.70	0.73	0.76	0.79	0.85	0.92	1.01	1.13	1.31
6.0	0.65	0.67	0.69	0.71	0.74	0.77	0.82	0.89	0.97	1.07	1.22
6.5	0.64	0.65	0.67	0.70	0.72	0.75	0.80	0.86	0.93	1.02	1.14
7.0	0.63	0.64	0.66	0.68	0.70	0.73	0.78	0.83	0.89	0.97	1.08
7.5	0.61	0.63	0.65	0.67	0.69	0.72	0.76	0.80	0.86	0.93	1.02
8.0	0.60	0.62	0.63	0.65	0.67	0.70	0.74	0.78	0.83	0.89	0.98
8.5	0.59	0.60	0.62	0.64	0.66	0.68	0.72	0.76	0.80	0.86	0.93
9.0	0.58	0.59	0.61	0.63	0.64	0.67	0.70	0.74	0.78	0.83	0.90
9.5	0.57	0.58	0.60	0.61	0.63	0.65	0.68	0.72	0.76	0.80	0.86
10.0	0.56	0.57	0.59	0.60	0.62	0.64	0.67	0.70	0.74	0.78	0.83
11.0	0.54	0.55	0.57	0.58	0.60	0.61	0.64	0.67	0.70	0.73	0.78
12.0	0.53	0.54	0.55	0.56	0.58	0.59	0.61	0.64	0.66	0.70	0.73
13.0	0.51	0.52	0.53	0.54	0.56	0.57	0.59	0.61	0.64	0.66	0.69
14.0	0.50	0.51	0.52	0.53	0.54	0.55	0.57	0.59	0.61	0.63	0.66
15.0	0.49	0.49	0.50	0.51	0.52	0.53	0.55	0.57	0.59	0.61	0.63
16.0	0.47	0.48	0.49	0.50	0.51	0.52	0.53	0.55	0.57	0.59	0.61
17.0	0.46	0.47	0.48	0.49	0.49	0.50	0.52	0.53	0.55	0.57	0.58
18.0	0.45	0.46	0.47	0.47	0.48	0.49	0.50	0.52	0.53	0.55	0.56
19.0	0.44	0.45	0.46	0.46	0.47	0.48	0.49	0.50	0.52	0.53	0.55
20.0	0.43	0.44	0.45	0.45	0.46	0.47	0.48	0.49	0.50	0.51	0.53

Note 1: For beams with *equal* flanges, $N = 0.5$; for beams with *unequal* flanges refer to clause 4.3.7.5 of BS 5950.

Note 2: v should be determined from the general formulae given in clause B.2.5 of BS 5950, on which this table is based: (a) for sections with *lipped* flanges (e.g. gantry girders composed of channel + universal beam); and (b) for intermediate values to the right of the stepped line in the table.

The factors m and n are interrelated as shown in BS 5950 Table 13, reproduced here as Table 5.8. From this table it can be seen that, when a beam is *not* loaded between points of lateral restraint, n is 1.0 and m should be obtained from BS 5950 Table 18. The value of m depends upon the ratio of the end moments at the points of restraint. If a beam *is* loaded between points of lateral restraint, m is 1.0 and n is obtained by reference

Table 5.8   Use of m and n factors for members of uniform section (BS 5950 Part 1 1990 Table 13)

Description		Members *not* subject to destabilizing loads*		Members subject to destabilizing loads*	
		m	n	m	n
Members loaded between adjacent lateral restraints	Sections with equal flanges	1.0	From Tables 15 and 16 of BS 5950	1.0	1.0
	Sections with unequal flanges	1.0	1.0	1.0	1.0
Members not loaded between adjacent lateral restraints	Sections with equal flanges	From Table 18 of BS 5950	1.0	1.0	1.0
	Sections with unequal flanges	1.0	1.0	1.0	1.0
Cantilevers without intermediate lateral restraints		1.0	1.0	1.0	` 1.0

* See clause 4.3.4 of BS 5950.

to BS 5950 Tables 15 and 16 (Table 16 is cross-referenced with Table 17). Its value depends upon the ratio of the end moments at the points of restraint and the ratio of the larger moment to the mid-span free moment.

Example 5.2

A simply supported steel beam spans 8 m and supports an ultimate central point load of 170 kN from secondary beams, as shown in Figure 5.11. In addition it carries an ultimate UDL of 9 kN resulting from its self-weight. If the beam is only restrained at the load position and the ends, determine a suitable grade 43 section.

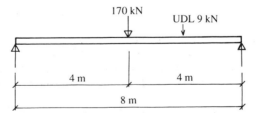

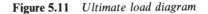

Figure 5.11   *Ultimate load diagram*

The maximum ultimate moment is given by

$$M_A = \frac{WL}{4} + \frac{WL}{8} = \frac{170 \times 8}{4} + \frac{9 \times 8}{8} = 340 + 9 = 349 \, \text{kN m}$$

Since the beam is laterally unrestrained it is necessary to select a trial section for checking: try $457 \times 152 \times 74 \, \text{kg/m UB}$ ($S_x = 1620 \, \text{cm}^3$). The moment capacity of this section when the beam is subject to low shear is given by $M_{cx} = p_y S_x$, where p_y is 265 N/mm^2 since T is greater than 16 mm. Thus

$$M_{cx} = p_y S_x = 265 \times 1620 \times 10^3 = 429.3 \times 10^6 \, N\,mm = 429.3 \, kN\,m > 349 \, kN\,m$$

This is adequate.

The lateral torsional buckling resistance is checked in the following manner:

$$\bar{M} = mM_A \leqslant M_b = p_b S_x$$

The self-weight UDL of 9 kN is relatively insignificant, and it is therefore satisfactory to consider the beam to be not loaded between restraints. By reference to Table 5.8, for members that are not subject to destabilizing loads, n is 1.0 and m should be obtained from BS 5950 Table 18.

The values of m in Table 18 depend upon β, which is the ratio of the smaller end moment to the larger end moment for the unrestrained length being considered. In this example the unrestrained length is the distance from a support to the central point load. The bending moment diagram for this length is shown in Figure 5.12.

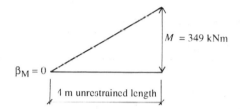

$\beta_M = 0$

$M = 349$ kNm

4 m unrestrained length

Figure 5.12 *Equivalent bending moment diagram for the unrestrained length*

It can be seen from this diagram that the end moment for a simply supported beam is zero. Hence

$$\beta - \frac{\text{smaller end moment}}{\text{larger end moment}} - \frac{0}{349} - 0$$

Therefore the value of m from BS 5950 Table 18 is 0.57.

It should be appreciated that if the central point load was from a column and there were no lateral beams at that point, then a destabilizing load condition would exist. In such a case both m and n, from Table 5.8, would be 1.0.

Equivalent uniform moment $\bar{M} = mM_A = 0.57 \times 349 = 198.93$ kN m

Buckling resistance moment $M_b = p_b S_x$

The bending strength p_b has to be obtained from Table 5.5 in relation to p_y and λ_{LT}. We have $p_y = 265$ N/mm^2 and

$$\lambda_{LT} = nuv\lambda$$

where $n = 1.0$, $u = 0.87$ from section tables, and $\lambda = L_E/r_y$. In this instance $L_E = 1.0L$ from Table 5.6, where L is the distance between restraints, and $r_y = 3.26$ cm $= 3.26 \times 10$ mm from section tables. Thus

$$\lambda = \frac{1.0 \times 4000}{3.26 \times 10} = 122.7$$

Now $x = 30$ from section tables. Hence $\lambda/x = 122.7/30 = 4.09$. and $v = 0.856$ by interpolation from Table 5.7. Hence

$$\lambda_{LT} = nuv\lambda = 1.0 \times 0.87 \times 0.856 \times 122.7 = 91.38$$

Therefore $p_b = 138.24 \, \text{N/mm}^2$ by interpolation from Table 5.5. Thus finally

$$M_b = p_b S_x = 138.24 \times 1620 \times 10^3$$
$$= 223.95 \times 10^6 \, \text{N mm} = 223.95 \, \text{kN m} > 198.93 \, \text{kN m}$$

That is, $\bar{M} < M_b$. Therefore the lateral torsional buckling resistance of the section is adequate. In conclusion:

Adopt $457 \times 152 \times 74 \, \text{kg/m UB}$.

The *Steelwork Design Guide* produced by the Steel Construction Institute also contains tables giving both the buckling resistance moment M_b and the moment capacity M_{cx} for the entire range of rolled sections. A typical example of a number of UB sections is reproduced here as Table 5.9. From the table it can be seen that for the $457 \times 152 \times 74 \, \text{kg/m UB}$ section that we have just checked, the relevant moment values are as follows:

$M_{cx} = 429 \, \text{kN m}$; and $M_b = 223 \, \text{kN m}$ when n is 1.0 and the effective length is 4.0 m. By using these tables the amount of calculation is significantly reduced, and they are therefore a particularly useful design aid for checking beams.

Example 5.3

If the beam in Example 5.2 were to be loaded between lateral restraints as shown in Figure 5.13, what size of grade 43 section would be required?

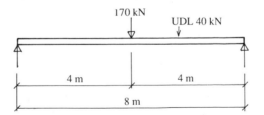

Figure 5.13   *Ultimate load diagram*

The maximum ultimate moment at mid-span is given by

$$M_A = \frac{WL}{4} + \frac{WL}{8} = \frac{170 \times 8}{4} + \frac{40 \times 8}{8} = 340 + 40 = 380 \, \text{kN m}$$

It is necessary to select a trial section for checking: try $457 \times 191 \times 82 \, \text{kg/m UB}$ ($S_x = 1830 \, \text{cm}^3$). Thus

$$M_{cx} = p_y S_x = 275 \times 1830 \times 10^3 = 503.25 \times 10^6 \, \text{N mm} = 503.25 \, \text{kN m} > 380 \, \text{kN m}$$

This is adequate.

Table 5.9 Universal beams subject to bending, steel grade 43: buckling resistance moment M_b(kN m) (abstracted from the *Steelwork Design Guide to BS 5950: Part 1*, published by the Steel Construction Institute)

Designation serial size: mass/metre and capacity	Slenderness correction factor n	Effective length L_E												
		2.0	2.5	3.0	3.5	4.0	4.5	5.0	6.0	7.0	8.0	9.0	10.0	11.0
$457 \times 191 \times 82$	0.4	503	503	503	503	503	503	496	472	451	431	413	395	379
$M_{cx} = 503$	0.6	503	503	500	478	457	436	417	379	346	317	291	269	249
Plastic	0.8	503	480	449	419	389	361	335	289	252	223	199	180	164
	1.0	478	437	396	357	321	289	261	217	184	159	140	126	114
$457 \times 191 \times 74$	0.4	456	456	456	456	456	456	446	424	403	384	366	349	333
$M_{cx} = 456$	0.6	456	456	451	430	410	391	372	337	305	277	253	232	214
Plastic	0.8	456	433	404	375	348	321	296	253	219	192	171	154	140
	1.0	431	393	355	319	285	255	230	189	159	137	120	107	96
$457 \times 191 \times 67$	0.4	404	404	404	404	404	402	391	370	350	332	314	298	283
$M_{cx} = 404$	0.6	404	404	397	378	359	341	323	290	260	234	212	194	178
Plastic	0.8	404	381	354	328	302	277	254	215	184	160	142	127	114
	1.0	380	345	310	277	246	219	195	159	132	113	98	87	78
$457 \times 152 \times 82$	0.4	477	477	477	477	475	462	450	427	407	388	370	0	0
$M_{cx} = 477$	0.6	477	471	447	424	402	381	362	327	297	272	250	0	0
Plastic	0.8	457	422	388	356	326	300	277	238	208	185	167	0	0
	1.0	416	370	327	290	257	231	208	174	149	131	116	0	0
$457 \times 152 \times 74$	0.4	429	429	429	429	423	411	399	377	357	339	322	0	0
$M_{cx} = 429$	0.6	429	421	398	376	355	335	317	284	256	232	212	0	0
Plastic	0.8	409	375	343	313	285	260	239	204	177	156	140	0	0
	1.0	371	328	288	252	223	198	178	147	125	109	97	0	0
$457 \times 152 \times 67$	0.4	396	396	396	396	384	372	360	338	318	299	283	0	0
$M_{ox} = 396$	0.6	396	383	361	339	318	299	280	247	220	198	179	0	0
Plastic	0.8	372	340	308	278	251	227	207	174	149	130	116	0	0
	1.0	336	294	255	221	193	170	152	124	105	90	79	0	0
$457 \times 152 \times 60$	0.4	352	352	352	351	339	328	317	296	276	259	243	0	0
$M_{cx} - 352$	0.6	352	340	319	299	280	261	244	213	188	167	151	0	0
Plastic	0.8	330	301	272	244	219	197	178	148	126	109	96	0	0
	1.0	298	260	224	193	168	147	130	105	87	75	66	0	0
$457 \times 152 \times 52$	0.4	300	300	300	295	284	274	263	243	225	208	194	0	0
$M_{cx} = 300$	0.6	300	286	267	249	231	214	198	170	148	130	116	0	0
Plastic	0.8	278	251	225	200	178	158	142	116	97	83	73	0	0
	1.0	249	215	183	156	134	116	102	81	67	57	50	0	0
$406 \times 178 \times 74$	0.4	412	412	412	412	412	412	404	386	369	354	339	326	313
$M_{cx} = 412$	0.6	412	412	405	387	370	354	338	309	283	260	240	223	208
Plastic	0.8	412	388	362	337	313	291	270	235	206	183	165	150	137
	1.0	385	351	317	286	257	232	210	175	150	131	116	105	95
$406 \times 178 \times 67$	0.4	371	371	371	371	371	370	360	343	327	312	298	285	273
$M_{cx} = 371$	0.6	371	371	363	346	330	314	299	271	246	225	206	190	176
Plastic	0.8	371	347	323	300	277	256	236	203	177	156	139	126	115
	1.0	345	313	282	252	226	202	182	151	128	111	97	87	79

M_b is obtained using an equivalent slenderness $= nuvL_e/r_y$.

Values have not been given for values of slenderness greater than 300.

The section classification given applies to members subject to bending only.

Check lateral torsional buckling:

$$\bar{M} = mM_A \leqslant M_b = p_b S_x$$

The magnitude of the UDL in this example is significant, and it will therefore be necessary to consider the beam to be loaded between lateral restraints. By reference to Table 5.8, for members not subject to destabilizing loads, m is 1.0 and n should be obtained from BS 5950 Table 16, which is cross-referenced with Table 17.

First we have

$$\bar{M} = mM_A = 1.0 \times 380 = 380 \, \text{kN m}$$

We now need to find M_b. The bending strength p_b has to be obtained from Table 5.5 in relation to p_y and λ_{LT}. We have $p_y = 275 \, \text{N/mm}^2$ and

$$\lambda_{LT} = nuv\lambda$$

The slenderness correction factor n is obtained from BS 5950 Table 16 in relation to the ratios γ and β for the length of beam between lateral restraints. In this instance that length would be from a support to the central point load. The ratios γ and β are obtained as follows. First, $\gamma = M/M_o$. The larger end moment $M = 380 \, \text{kN m}$. M_o is the mid-span moment on a simply supported span equal to the unrestrained length, that is $M_o = WL/8$. The UDL on unrestrained length is $W = 40/2 = 20 \, \text{kN}$, and the unrestrained length $L = \text{span}/2 = 4 \, \text{m}$. Hence

$$M_o = \frac{WL}{8} = \frac{20 \times 4}{8} = 10 \, \text{kNm}$$

The equivalent bending moment diagram, for the unrestrained length, corresponding to these values is shown in Figure 5.14. Thus

$$\gamma = \frac{M}{M_o} = \frac{380}{10} = 38$$

Secondly,

$$\beta = \frac{\text{smaller end moment}}{\text{larger end moment}} = \frac{0}{380} = 0$$

Therefore, by interpolation from BS 5950 Table 16, $n = 0.782$.
From section tables, $u = 0.877$.

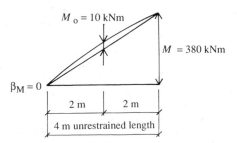

Figure 5.14 *Equivalent bending moment diagram for the unrestrained length*

Next $\lambda = L_E/r_y$, where $L_E = 1.0L$ in this instance from Table 5.6; L is the distance between restraints; and $r_y = 4.23$ cm from section tables, that is 4.23×10 mm. Thus

$$\lambda = \frac{L_E}{r_y} = \frac{1.0 \times 4000}{4.23 \times 10} = 94.56$$

Here $x = 30.9$ from section tables. Therefore $\lambda/x = 94.56/30.9 = 3.06$, and so $v = 0.91$ from Table 5.7.

Finally, therefore,

$$\lambda_{LT} = nuv\lambda = 0.782 \times 0.877 \times 0.91 \times 94.56 = 59$$

Using the values of p_y and λ_{LT}, $p_b = 215.6$ N/mm^2 by interpolation from Table 5.5. In conclusion,

$$M_b = p_b S_x = 215.6 \times 1830 \times 10^3 = 394.5 \times 10^6 \text{ N mm} = 394.5 \text{ kN m} > 380 \text{ kN m}$$

Thus $\bar{M} < M_b$, and therefore the lateral torsional buckling resistance of the section is adequate.

Adopt $457 \times 191 \times 82$ kg/m UB.

The M_{cx} and M_b values that we have calculated may be compared with those tabulated by the Steel Construction Institute for a $457 \times 191 \times 82$ kg/m UB. From Table 5.9, $M_{cx} = 503$ kN m, and $M_b = 389$ kN m when n is 0.8 and the effective length is 4.0.

Laterally unrestrained beams, conservative approach

The suitability of laterally unrestrained UB, UC and RSJ sections may be checked, if desired, using a conservative approach. It should be appreciated that being conservative the design will not be as economic as that given by the rigorous approach; consequently beam sections that are proved to be adequate using the rigorous approach may occasionally prove inadequate using the conservative approach. However, it does have the advantage that members either loaded or unloaded between restraints are checked using one expression.

In the conservative approach the maximum moment M_x occurring between lateral restraints must not exceed the buckling resistance moment M_b:

$$M_x \leqslant M_b$$

The buckling resistance moment is given by the expression

$$M_b = p_b S_x$$

For the conservative approach, p_b is obtained from the appropriate part of Table 19a–d of BS 5950 in relation to λ and x, the choice depending on the design strength p_y of the steel.

Loads occurring between restraints may be taken into account by multiplying the effective length by a slenderness correction factor n obtained either from BS 5950 Table 13 (reproduced earlier as Table 5.8) or alternatively from BS 5950 Table 20, except for destabilizing loads when it should be taken as 1.0. It is important to understand that the reactions shown on the diagrams in Table 20 are the lateral restraints and not just the beam supports. Therefore for a simply supported beam with a central point load providing lateral restraint, the relevant Table 20 diagram would be as shown in Figure 5.15. The corresponding value of n would then be 0.77.

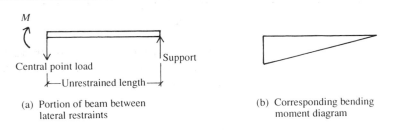

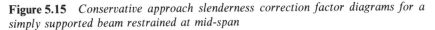

(a) Portion of beam between lateral restraints

(b) Corresponding bending moment diagram

Figure 5.15 *Conservative approach slenderness correction factor diagrams for a simply supported beam restrained at mid-span*

Thus the minor axis slenderness ratio is given by

$$\lambda = \frac{nL_E}{r_y}$$

where n is the slenderness correction factor either from BS 5950 Table 13 or Table 20, L_E is the effective unrestrained length of the beam, and r_y is the radius of gyration of the section about its minor axis, found from section tables. The torsional index x of the section is taken from section tables.

For those who are familiar with BS 449, this approach is similar to the use of Table 3 in that standard, which was related to the l/r and D/T ratios of the section.

Example 5.4

Check the beam section selected in Example 5.3, using the conservative approach.

The maximum ultimate moment $M_x = 380\,\text{kN}\,\text{n}$ at midspan. Check $457 \times 191 \times 82\,\text{kg/m}$ UB ($S_x = 1830\,\text{cm}^3$). $T = 16\,\text{mm}$; hence $p_y = 275\,\text{N/mm}^2$. Thus

$$M_{cx} = p_y S_x = 275 \times 1830 \times 10^3 = 503.25 \times 10^6\,\text{N mm} = 503.25\,\text{kN m} > 380\,\text{kN m}$$

This is satisfactory:

Check lateral torsional buckling, that is show

$$M_x \leqslant M_b = p_b S_x$$

For the conservative approach, p_b is obtained from BS 5950 Table 19b when p_y is 275 N/mm², using λ and x. The slenderness correction factor n obtained from BS 5950 Table 20 is 0.77. Then

$$\lambda = \frac{nL_E}{r_y} = \frac{0.77 \times 4000}{4.23 \times 10} = 72.8$$

Now $x = 30.9$. Thus $p_b = 210\,\text{N/mm}^2$ by interpolation from BS 5950 Table 19b. So

$$M_b = p_b S_x = 210 \times 1830 \times 10^3 = 384.3 \times 10^6\,\text{N mm} = 384.3\,\text{kN m} > 380\,\text{kN m}$$

Therefore $M_x < M_b$, and so the lateral torsional buckling resistance of the section is adequate.

5.10.4 Shear ULS

The shear resistance of a beam is checked by ensuring that the ultimate shear force F_v does not exceed the shear capacity P_v of the section at the point under consideration:

$$F_v \leqslant P_v$$

where

F_v ultimate shear force at point under consideration

P_v shear capacity of section: $P_v = 0.6 p_y A_v$

p_y design strength of steel, given in Table 5.1.

A_v area of section resisting shear: $A_v = tD$ for rolled sections, as shown in Figure 5.16

t total web thickness, from section tables

D overall depth of section, from section tables

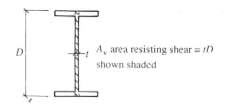

Figure 5.16 *Area of a rolled section resisting shear*

It is recommended in BS 5950 that the combination of maximum moment and coexistent shear, and the combination of maximum shear and coexistent moment, should be checked. The moment capacity of plastic and compact beam sections is reduced when high shear loads occur. A high shear load is said to exist when the ultimate shear force exceeds 0.6 times the shear capacity of the section, that is when $F_v > 0.6 P_v$. However, as mentioned in Example 5.1, this is not usually a problem except for heavily loaded short span beams.

When the depth to thickness ratio d/t of a web exceeds 63ε, where $\varepsilon = (275/p_y)^{1/2}$ as previously referred to in Table 5.4, the web should be checked for shear buckling. This does not apply to any of the standard rolled sections that are available, but it may apply to plate girders made with thin plates.

It should be appreciated that, if necessary, the web of a beam may be strengthened locally to resist shear by the introduction of stiffeners, designed in accordance with the recommendations given in BS 5950.

Example 5.5

Check the shear capacity of the beam that was designed for bending in Example 5.1. The loading, shear force and bending moment diagrams for the beam are shown in Figure 5.17.

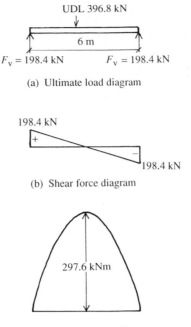

UDL 396.8 kN

6 m

$F_v = 198.4$ kN $F_v = 198.4$ kN

(a) Ultimate load diagram

198.4 kN

+

−

198.4 kN

(b) Shear force diagram

297.6 kNm

(c)   Bending moment diagram

Figure 5.17 *Beam diagrams for ultimate loads*

The section selected to resist bending was a $457 \times 152 \times 60$ kg/m UB, for which the relevant properties for checking shear, from Table 5.2, are $t = 8.0$ mm and $D = 454.7$ mm. Beam sections should normally be checked for the combination of maximum moment and coexistent shear, and the combination of maximum shear and coexistent moment. However, since the beam in this instance only carried a UDL the shear is zero at the point of maximum moment. Therefore it will only be necessary to check the section at the support where the maximum shear occurs and the coexistent moment is zero.

Ultimate shear at support $F_v = 198.4$ kN

Shear capacity of section $P_v = 0.6 p_y A_v = 0.6 p_y t D$

$$= 0.6 \times 275 \times 8 \times 454.7 = 600\,204 \text{ N}$$
$$= 600 \text{ kN} > 198 \text{ kN}$$

That is $F_v < P_v$, and therefore the section is adequate in shear.

Example 5.6

Check the shear capacity of the beam that was designed for bending in Example 5.2. The loading, shear force and bending moment diagrams for the beam are shown in Figure 5.18.

The section selected to resist bending was a $457 \times 152 \times 74$ kg/m UB, for which the relevant properties for checking shear, from Table 5.2, are $t = 9.9$ mm and $D = 461.3$ mm. In addition it should be noted that the flange thickness T of this section

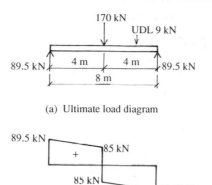

(a) Ultimate load diagram

(b) Shear force diagram

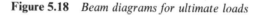

(c) Bending moment diagram

Figure 5.18   *Beam diagrams for ultimate loads*

is greater than 16 mm, and therefore the reduced p_y value of 265 N/mm² should be used in the calculations.

This beam will be checked for the combination of maximum moment and co-existent shear, and the combination of maximum shear and co-existent moment.

Maximum moment and coexistent shear at midspan

Ultimate shear at midspan $F_v = 85$ kN; $M = 349$ kN m

Shear capacity of section $P_v = 0.6 p_y t D$

$$= 0.6 \times 265 \times 9.9 \times 461.3 = 726\,132\,\text{N}$$
$$= 726\,\text{kN} > 85\,\text{kN}$$

Furthermore $0.6 P_v = 0.6 \times 726 = 435.6$ kN. Therefore

$$F_v = 85\,\text{kN} < 0.6 P_v = 435.6\,\text{kN}$$

Hence the shear load is low and no reduction to the moment capacity calculated earlier is necessary because of shear.

Maximum shear and co-existent moment

Ultimate shear at support $F_v = 89.5$ kN $M = 0$

Shear capacity of section $P_v = 726$ kN > 89.5 kN

That is $F_v < P_v$, and therefore the section is adequate in shear.

5.10.5   Deflection SLS

The deflection limits for steel beams are given in BS 5950 Table 5. For beams carrying plaster or other brittle finish the limit is span/360, and for all other beams is span/200. That is,

$$\text{Permissible deflection } \delta_p = \frac{\text{span}}{360} \quad \text{or} \quad \frac{\text{span}}{200}$$

It should be appreciated that these are only recommended limits and in certain circumstances more stringent limits may be appropriate. For example the deflection of beams supporting glazing or door gear may be critical to the performance of such items, in which case a limit of span/500 may be more realistic.

The actual deflection produced by the unfactored imposed loads alone should be compared with these limits. This is calculated using the formula relevant to the applied loading. For example,

$$\text{Actual deflection due to a UDL } \delta_a = \frac{5}{384} \frac{WL^3}{EI}$$

I is cm^4 to m^4 ×10^4

$$\text{Actual deflection due to a central point load } \delta_a = \frac{1}{48} \frac{WL^3}{EI}$$

where, in relation to steel sections, $E = 205 \, \text{kN/mm}^2 = 205 \times 10^3 \, \text{N/mm}^2$, and I is the second moment of area of the section about its major x–x axis, found from section tables. That is,

$$\delta_a \leqslant \delta_p$$

Example 5.7

Check the deflection of the beam that was designed for bending in Example 5.3 if the unfactored imposed loads are as shown in Figure 5.19.

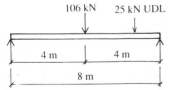

106 kN   25 kN UDL

4 m   4 m

8 m

Figure 5.19   *Unfactored imposed loads*

The section selected to resist bending was a $457 \times 191 \times 82 \, \text{kg/m}$ UB, for which the second moment of area I_x is $37\,100 \, \text{cm}^4$. The deflection limit is given by

$$\delta_p = \frac{\text{span}}{360} = \frac{8000}{360} = 22.22 \, \text{mm}$$

The actual deflection is

$$\delta_a = \frac{5}{384} \frac{WL^3}{EI} + \frac{1}{48} \frac{WL^3}{EI}$$

$$= \frac{5}{384} \times \frac{25 \times 10^3 \times 8000^3}{205 \times 10^3 \times 37\,100 \times 10^4}$$

$$\quad + \frac{1}{48} \times \frac{106 \times 10^3 \times 8000^3}{205 \times 10^3 \times 37\,100 \times 10^4}$$

$$= 2.19 + 14.87 = 17.06 \, \text{mm} < 22.22 \, \text{mm}$$

That is $\delta_a < \delta_p$, and therefore the section is adequate in deflection.

5.10.6   Web buckling resistance

When a concentrated load, such as the reaction, is transmitted through the flange of a beam to the web, it may cause the web to buckle. In resisting such buckling the web of the beam behaves as a strut. The length of web concerned is determined on the assumption that the load is dispersed at 45° from the edge of stiff bearing to the neutral axis of the beam, as shown in Figure 5.20. The buckling resistance P_w of the un-stiffened web is calculated from the following expression:

$$P_w = (b_1 + n_1)tp_c$$

where

b_1 stiff bearing length

n_1 length obtained by dispersion through half the depth of the section

t web thickness, from section tables

p_c compressive strength of the steel

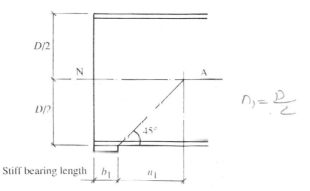

$$n_1 = \frac{D}{2}$$

Figure 5.20   *Web buckling resistance: load dispersal*

The compressive strength p_c of the steel should be obtained from BS 5950 Table 27c in relation to the ultimate design strength of the steel and the web slenderness λ.

When the beam flange through which the load is applied is restrained against rotation relative to the web and against lateral movement relative to the other flange, then the slenderness is given by the following expression from BS 5950:

$$\lambda = 2.5 \frac{d}{t}$$

Should these conditions not apply, then the slenderness may conservatively be obtained using the following expression:

$$\lambda = 3.46 \frac{d}{t}$$

Example 5.8

Check the web buckling capacity of the beam that was designed for bending in Example 5.1. It may be assumed that the beam is supported on a stiff bearing length of 75 mm as indicated in Figure 5.21.

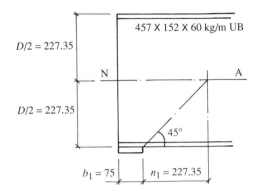

Figure 5.21   *Web buckling check dimensions*

From the loading diagram for this beam, shown in Figure 5.17, the maximum ultimate reaction is 198.4 kN.

The section selected to resist bending was a $457 \times 152 \times 60$ kg/m UB, for which the relevant properties for checking web buckling, from Table 5.2, are as follows:

$$D = 454.7 \, \text{mm} \qquad \frac{D}{2} = \frac{454.7}{2} = 227.35 \, \text{mm}$$

$$\frac{d}{t} = 51.00 \qquad t = 8.0 \, \text{mm}$$

With both flanges restrained,

$$\lambda = 2.5 \frac{d}{t} = 2.5 \times 51 = 127.5$$

Also $p_y = 275 \, \text{N/mm}^2$. Thus by interpolation from BS 5950 Table 27c, $p_c = 88.5$ N/mm^2.

The stiff bearing length $b_1 = 75$ mm, and $n_1 = D/2 = 227.35$ mm. Hence

$$P_w = (b_1 + n_1)tp_c = (75 + 227.35)8 \times 88.5 = 214\,064 \, \text{N} = 214 \, \text{kN} > 198.4 \, \text{kN}$$

Thus the buckling resistance of the unstiffened web is greater than the maximum reaction, and therefore the web does not require stiffening to resist buckling.

5.10.7   Web bearing resistance

The web bearing resistance of a beam is the ability of its web to resist crushing induced by concentrated loads such as the reactions. These are considered to be dispersed through the flange at a slope of 1:2.5 to the point where the beam flange joins the web, that being the critical position for web bearing (see Figure 5.22).

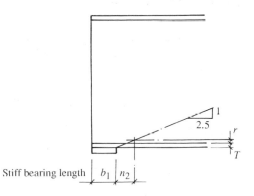

Figure 5.22 *Web bearing resistance: load dispersal*

The ultimate web bearing capacity P_{crip} of a beam is given by the following expression:

$$P_{crip} = (b_1 + n_2)tp_{yw}$$

where

b_1 stiff bearing length

n_2 length obtained by dispersion at a slope of 1:2.5 through the flange to the flange to web connection: $n_2 = 2.5(r + T)$

r root radius of the beam, from section tables

T beam flange thickness, from section tables

p_{yw} design strength of the web: $p_{yw} = p_y$

Example 5.9

Check the web bearing capacity of the beam that was designed for bending in Example 5.1. It may be assumed that the beam is supported on a stiff bearing length of 75 mm, as indicated in Figure 5.23.

From the loading diagram for this beam shown in Figure 5.17, the maximum ultimate reaction is 198.4 kN.

The section selected to resist bending was a $457 \times 152 \times 60$ kg/m UB, for which the relevant properties for checking web bearing, from Table 5.2, are as follows:

$$r = 10.2 \text{ mm} \qquad T = 13.3 \text{ mm} \qquad t = 8.0 \text{ mm}$$
$$\text{Stiff bearing length } b_1 = 75 \text{ mm}$$
$$n_2 = 2.5(r + T) = 2.5(10.2 + 13.3) = 58.75 \text{ mm}$$

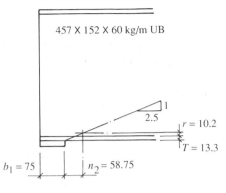

457 X 152 X 60 kg/m UB

2.5

$r = 10.2$

$T = 13.3$

$b_1 = 75$ $n_2 = 58.75$

Figure 5.23   *Web bearing check dimensions*

Hence

$$P_{\text{crip}} = (b_1 + n_2)tp_{yw} = (75 + 58.75)8 \times 275 = 294\,250\,\text{N} = 294\,\text{kN} > 198.4\,\text{kN}$$

Thus the bearing resistance of the unstiffened web is greater than the maximum reaction, and therefore the web does not require stiffening to resist crushing due to bearing.

5.10.8   Design summary for steel beams

Having examined the various aspects that can influence the design of steel beams, the general procedure when using grade 43 rolled sections may be summarized as follows.

Bending

LOADS FACTORED

(a) Decide if the beam will be laterally restrained or laterally unrestrained.

(b) If the beam is laterally restrained, ensure that the moment capacity M_{cx} of the section is greater than the applied ultimate moment M_u:

$$M_{cx} = p_y S_x \geqslant M_u$$

(c) If the beam is laterally unrestrained, the lateral torsional buckling resistance of the section will have to be checked. This may be done using a rigorous approach or a conservative approach. In both methods, account should be taken of any loading between restraints.

(i) Using the rigorous approach, ensure that the applied equivalent uniform moment $\bar{M}$ is less than the buckling resistance moment M_b of the section:

$$\bar{M} = mM_A \leqslant M_b = p_b S_x$$

(ii) Using the conservative approach, ensure that the maximum moment M_x occurring between lateral restraints does not exceed the buckling resistance moment M_b of the section:

$$M_x \leqslant M_b = p_b S_x$$

$P_b < P_y$

Shear

Both the combination of maximum moment and coexistent shear, and the combination of maximum shear and coexistent moment, should be checked.

The shear resistance of a beam is checked by ensuring that the ultimate shear force F_v does not exceed the shear capacity P_v of the section at the point under consideration:

$$F_v \leqslant P_v$$

It should be noted that the moment capacity of plastic and compact beam sections must be reduced when high shear loads occur. However, this is not usually a problem except for heavily loaded short span beams.

A high shear load condition exists when

$$F_v > 0.6P_v$$

Deflection

The deflection requirement of a beam is checked by comparing the actual deflection produced by the unfactored imposed loads with the recommended limits given in BS 5950 Table 5:

$$\text{Actual deflection} < \text{recommended deflection limit}$$

Web buckling

The web buckling resistance of an unstiffened web must be greater than any concentrated load that may be applied.

Web bearing

The web bearing resistance of an unstiffened web must be greater than any concentrated load that may be applied.

It should be appreciated that the requirements for web buckling and bearing are not usually critical under normal loading conditions. Furthermore, they can if necessary be catered for by the inclusion of suitably designed web stiffeners.

Before leaving the topic of beams, let us look at a further example illustrating the complete design of a laterally unrestrained beam using the rigorous approach.

Example 5.10

The simply supported beam shown in Figure 5.24 is laterally restrained at the ends and at the points of load application. For the loads given below, determine the size of grade 43 section required.

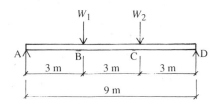

Figure 5.24 *Simply supported beam*

Specified dead loads:

Point load $W_{1d} = 30\,kN$; point load $W_{2d} = 20\,kN$

Self-weight $= 1\,kN/m$; SW UDL $= 1 \times 9 = 9\,kN$

Specified imposed loads:

Point load $W_{1i} = 50\,kN$; point load $W_{2i} = 30\,kN$

Ultimate design loads:

$W_1 = \gamma_f W_{1d} + \gamma_f W_{1i} = 1.4 \times 30 + 1.6 \times 50 = 42 + 80 = 122\,kN$

$W_2 = \gamma_f W_{2d} + \gamma_f W_{2i} = 1.4 \times 20 + 1.6 \times 30 = 28 + 48 = 76\,kN$

SW UDL $= 1.4 \times 9 = 12.6\,kN$

The ultimate design load diagram and the corresponding shear force and bending moment diagrams are shown in Figure 5.25. Since the loading is not symmetrical, the reactions and moments are calculated from first principles.

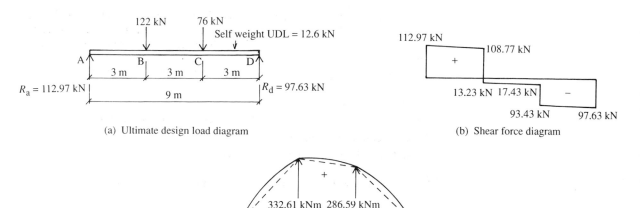

(a) Ultimate design load diagram

(b) Shear force diagram

(c) Bending moment diagram

Figure 5.25 *Beam diagrams for ultimate loads*

For the reactions, take moments about D:

$$9R_a = (122 \times 6) + (76 \times 3) + (12.6 \times 4.5) = 732 + 228 + 56.7 = 1016.7$$
$$R_a = 1016.7/9 = 112.97\,kN$$
$$R_d = (122 + 76 + 12.6) - 112.97 = 97.63\,kN$$

Ultimate moment at B $= 112.97 \times 3 - \dfrac{12.6}{9} \times \dfrac{3^2}{2} = 332.61\,kN\,m$

Ultimate moment at C $= 97.63 \times 3 - \dfrac{12.6}{9} \times \dfrac{3^2}{2} = 286.59\,kN\,m$

Since the beam is not fully restrained laterally, the buckling resistance moment M_b of the section needs to be checked in comparison with the applied equivalent uniform moment $\bar{M}$ to ensure that

$$\bar{M} = mM_A \leqslant M_b - p_b S_x$$

By reference to the bending moment diagram shown in Figure 5.25c, the critical unrestrained length will be BC where the maximum moment occurs.

The self-weight UDL is relatively insignificant and it is therefore satisfactory to consider the beam to be unloaded between restraints. Hence n is 1.0 and m is obtained from BS 5950 Table 18. We have

$$\beta = \frac{\text{smaller end moment}}{\text{larger end moment}} = \frac{M \text{ at C}}{M \text{ at B}} = \frac{286.59}{332.61} = 0.86$$

Therefore by interpolation from Table 18, $m = 0.93$. The maximum moment on length BC is $M_A = M$ at B $= 332.61\,kN\,m$. Hence

$$M = mM_A = 0.93 \times 332.61 = 309.33\,kN\,m$$

The effective length L_E of BC is 3.0 m.

By reference to Table 5.9, reproduced from the Steel Construction Institute design guide, a $457 \times 191 \times 74\,kg/m$ UB has a buckling resistance moment of $355\,kN\,m$ when n is 1 and L_E is 3.0 m. Therefore let us check this section in bending, shear and deflection.

The relevant properties for the section from tables are as follows:

Plastic modulus $S_x = 1660\,cm^3$

$D = 457.2\,mm$ $\qquad t = 9.1\,mm$ $\qquad T = 14.5\,mm$ $\qquad d = 407.9\,mm$

Section classification: plastic

Since $T = 14.5\,mm < 16\,mm$, $p_y = 275\,N/mm^2$.

Check the section for combined moment and shear as follows.

Maximum moment and coexistent shear at B

Ultimate shear at B is $F_v = 108.77\,kN$; $M = 332.61\,kN\,m$

Shear capacity of section is $P_v = 0.6 p_y t D$
$$= 0.6 \times 275 \times 9.1 \times 457.2 = 686\,486\,N$$
$$= 686\,kN > 108.77\,kN$$

This is satisfactory. Furthermore, $0.6P_v = 0.6 \times 686 = 412\,\text{kN}$. Therefore

$$F_v = 108.77\,\text{kN} < 0.6P_v = 412\,\text{kN}$$

Hence the shear load is low and the moment capacity is as follows:

$$M_{cx} = p_y S_x = 275 \times 1660 \times 10^3 = 456.5 \times 10^6 \,\text{N}\,\text{mm} = 456.5\,\text{kN}\,\text{m} > 332.61\,\text{kN}\,\text{m}$$

Maximum shear and coexistent moment at A

Ultimate shear at A is $F_v = 112.97\,\text{kN}$; $M = 0$

P_v is again $686\,\text{kN} > 112.97\,\text{kN}$

Buckling resistance

The lateral torsional buckling resistance has already been satisfied by selecting a section from Table 5.9 with a buckling resistance moment M_b greater than the equivalent uniform moment $\bar{M}$. However, the method of calculating the buckling resistance moment in accordance with BS 5950 will be included here for reference.

The buckling resistance moment of the section is given by

$$M_b = p_b S_x$$

The bending strength p_b is obtained from Table 5.5 in relation to p_y and λ_{LT}. We have $p_y = 275\,\text{N/mm}^2$ and

$$\lambda_{LT} = nuv\lambda$$

Now $n = 1.0$, and $u = 0.876$ from section tables. Next $\lambda = L_E/r_y$, where $L_E = 1.0L$ in this instance from Table 5.6; L is the distance BC between restraints; and $r_y = 4.19\,\text{cm} = 4.19 \times 10\,\text{mm}$ from section tables. Thus

$$\lambda = \frac{L_E}{r_y} = \frac{1.0 \times 3000}{4.19 \times 10} = 71.6$$

Here $x = 33.9$ from section tables. Thus $\lambda/x = 71.6/33.9 = 2.11$, and so $v = 0.95$ by interpolation from Table 5.7.

Finally, therefore,

$$\lambda_{LT} = nuv\lambda = 1.0 \times 0.876 \times 0.95 \times 71.6 = 59.6$$

Using the values of p_y and λ_{LT}, $p_b = 214\,\text{N/mm}^2$ from Table 5.5. In conclusion,

$$M_b = p_b S_x = 214 \times 1660 \times 10^3 = 355.2 \times 10^6 \,\text{N}\,\text{mm} = 355.2\,\text{kN}\,\text{m} > 309.33\,\text{kN}\,\text{m}$$

Thus $\bar{M} < M_b$, and therefore the lateral torsional buckling resistance of the section is adequate.

Deflection

Since the loading on this beam is not symmetrical, the calculations needed to determine the actual deflection are quite complex. A simpler approach is to calculate the deflection due to an equivalent UDL and compare it with the permitted limit of span/360. If this proves that the section is adequate then there would be no need to resort to more exact calculations.

The deflection should be based upon the unfactored imposed loads alone. These and the resulting shear and bending moment diagrams are shown in Figure 5.26. By equating the maximum bending moment of 129 kN m to the expression for the bending moment due to a UDL, an equivalent UDL can be calculated:

$$129 = \frac{WL}{8}$$

$$W = \frac{8 \times 129}{L} = \frac{8 \times 129}{9} = 115 \, \text{kN}$$

This equivalent UDL of 115 kN may be substituted in the expression for the deflection of a simply supported beam:

Actual deflection $\delta_a = \dfrac{5}{384} \dfrac{WL^3}{EI} = \dfrac{5}{384} \times \dfrac{115 \times 10^3 \times 9000^3}{205 \times 10^3 \times 33\,400 \times 10^4} = 15.94 \, \text{mm}$

Deflection limit $\delta_p = \dfrac{\text{span}}{360} = \dfrac{9000}{360} = 25 \, \text{mm}$

Thus $\delta_a < \delta_p$, and the beam is satisfactory in deflection.

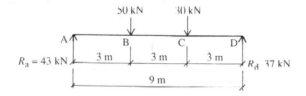

(a) Unfactored imposed load diagram

43 kN

+

7 kN

—

37 kN

(b) Shear force diagram

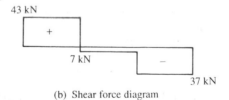

129 kNm   111 kNm

(c) Bending moment diagram

Figure 5.26 *Beam diagrams for unfactored imposed loads*

Web buckling and bearing

The web buckling and bearing requirements are not critical and therefore the calculations for these will be omitted.

Conclusion

That completes the check on the section, which has been shown to be adequate in bending, shear and deflection. Thus:

Adopt $457 \times 191 \times 74$ kg/m UB.

5.11 Fabricated beams

In situations where standard rolled sections are found to be inadequate, consideration should be given to the following fabricated alternatives.

Compound beams

The strength of standard rolled sections can be increased by the addition of reinforcing plates welded to the flanges. Beams strengthened in this way are called compound beams. Examples are shown in Figure 5.27.

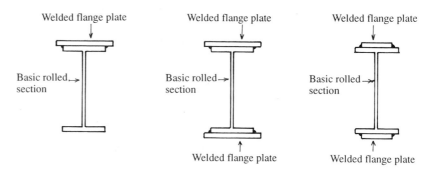

Figure 5.27   *Examples of compound beams*

Castellated beams

Standard rolled sections can be converted by cutting and welding into much deeper sections known as castellated beams. They offer a relatively simple method of increasing the strength of a section without increasing its weight.

To form a castellated beam, the basic rolled section is first flame cut along its web to a prescribed profile as shown in Figure 5.28a. Then the resulting two halves are rejoined by welding to form the castellated beam shown in Figure 5.28b. The finished section is stronger in bending than the original but the shear strength is less. However, this usually only affects heavily loaded short span beams, and may be overcome where necessary by welding fitted plates into the end castellations as shown in Figure 5.28c.

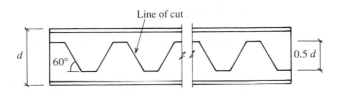

(a) Web of basic rolled section cut to prescribed profile

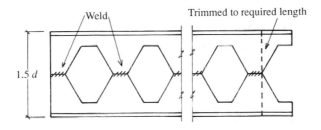

(b) Two-halves re-joined to form castellated beam

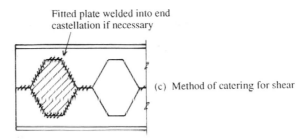

(c) Method of catering for shear

Figure 5.28 *Castellated beams*

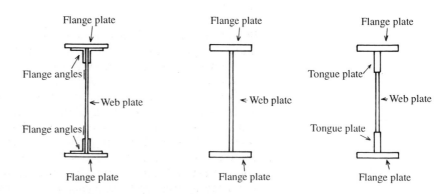

Figure 5.29 *Examples of plate girders*

Plate girders

Plate girders are used occasionally in buildings where heavy loads or long spans dictate, but more often they are used for bridges. They are formed from steel plates, sometimes in conjunction with angles, which are welded or bolted together to form I-sections. Three of the most common forms are illustrated in Figure 5.29.

Whilst plate girders can theoretically be made to any size, their depth for practical reasons should usually be between span/8 and span/12.

Lattice girders

Lattice girders are a framework of individual members bolted or welded together to form an open web beam. Two types of lattice girder commonly encountered are illustrated in Figure 5.30; they are the N-girder and the Warren girder. In comparison with the structural behaviour of beams, the top and bottom booms of a lattice girder resist bending and the internal members resist shear.

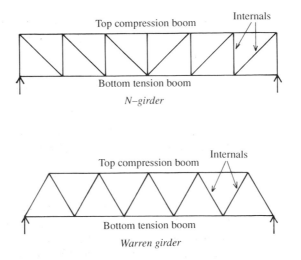

Figure 5.30 *Examples of lattice girders*

Generally their economical depth is between span/10 and span/15. Exceptions are short span heavily loaded girders, for which the depth may equal span/6, and long span lightly loaded roof girders, for which a depth of span/20 may suffice.

5.12 Columns

A steel column may be subject to direct compression alone, where the load is applied axially, or subject to a combination of compressive loading and bending due to the load being applied eccentrically to the member axes. It may also be subject to horizontal bending induced by lateral wind loading. However, the effect of wind loading on individual structural elements is not being considered in this manual.

Guidance for the design of axially loaded columns and axially loaded columns with moments is given in BS 5950 Part 1. The procedure for dealing with columns subject to axial load alone is first explained. This is then extended to include the interaction between compression and bending. Separate guidance is also given for the design of concrete cased columns and baseplates for columns.

The design of steel columns in this manual will therefore be considered under the following headings:

(a) Axially loaded columns

(b) Axially loaded columns with moments

(c) Cased columns

(d) Column base plates.

5.12.1 Axially loaded columns

A column supporting an axial load is subjected to direct compression. The compression resistance P_c of a column is given by the following expression:

$$P_c = A_g p_c$$

where A_g is the gross sectional area and p_c is the compressive strength. To ensure that a particular steel column is adequate, its compression resistance must be equal to or greater than the ultimate axial load F:

$$P_c \geqslant F$$

A steel column, because of its slender nature, will tend to buckle laterally under the influence of the applied compression. Therefore the compressive strength p_c is reduced to take account of the slenderness of the column. The slenderness λ of an axially loaded column is given by the following expression:

$$\lambda = \frac{L_E}{r}$$

where L_E is the effective length of the column and r is the radius of gyration of the section about the relevant axis, found from section tables.

The maximum slenderness of steel columns carrying dead and imposed loads is limited to 180. Values greater than this limit indicate that a larger section size is required.

Guidance on the nominal effective lengths to be adopted, taking end restraint into consideration, is given in BS 5950 Table 24. Additional guidance in relation to columns in certain single storey buildings and those forming part of a rigid frame are given in Appendices D and E respectively of the standard. The main effective length requirements for single storey steel sections are summarized here in Table 5.10.

Table 5.10   Effective length of steel columns

End condition	Effective length L_E
Restrained at both ends in position and direction	0.7 L
Restrained at both ends in position and one end in direction	0.85 L
Restrained at both ends in position but not in direction	1.0 L
Restrained at one end in position and in direction and at the other end in direction but not in position	1.5 L
Restrained at one end in position and in direction and free at the other end	2.0 L

The compressive strength p_c depends on the slenderness λ and the design strength of the steel p_y, or on a reduced design strength if the section is classified as slender. It was mentioned previously with respect to beams that the web and flanges of steel sections are comparatively slender in relation to their depth and breadth. Consequently the compressive force acting on a column could also cause local buckling of the web or flange before the full plastic stress is developed. This situation is avoided by reducing the stress capacity of the columns in relation to its section classification.

The column designs contained in this manual will be related to the use of UC sections, which are defined as H-sections in BS 5950. Since it can be shown that all UC sections are classified as at least semi-compact when used as axially loaded columns, no reduction in the design strength p_y because of local buckling will be necessary.

The value of the compressive strength p_c in relation to the slenderness λ and the design strength p_y is obtained from strut tables given in BS 5950 as Table 27a–d. The specific table to use is indicated in Table 25 of the standard relative to the type of section employed. With respect to UC sections, the particular strut table to use is given here as Table 5.11.

Table 5.11   Selection of BS 5950 strut table

Section type	Thickness	Buckling axis	
		x–x	y–y
Rolled H-section	Up to 40 mm	Table 27b	Table 27c
	Over 40 mm	Table 27c	Table 27d

5.12.2 Design summary for axially loaded steel columns

The general procedure for the design of axially loaded columns, using grade 43 UC sections, may be summarized as follows:

(a) Calculate the ultimate axial load F applied to the column.
(b) Determine the effective length L_E from the guidance given in Table 5.10.
(c) Select a trial section.
(d) Calculate the slenderness λ from L_E/r and ensure that it is not greater than 180.
(e) Using the slenderness λ and steel design strength p_y, obtain the compression strength p_c of the column from Table 27a–d of BS 5950.
(f) Calculate the compression resistance P_c of the column from the expression $P_c = A_g p_c$, where A_g is the gross sectional area of the column.
(g) Finally, check that the compression resistance P_c is equal to or greater than the ultimate axial load F.

Example 5.11

Design a suitable grade 43 UC column to support the ultimate axial load shown in Figure 5.31. The column is restrained in position at both ends but not restrained in direction.

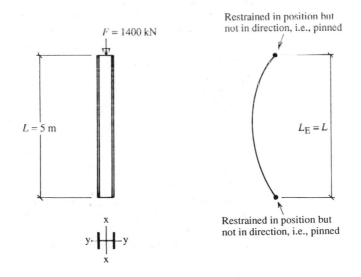

Figure 5.31 *Column load and effective lengths*

Ultimate axial load $F = 1400\,\text{kN}$
Effective length $L_E = 1.0L = 5000\,\text{mm}$

It is first necessary to assume a trial section for checking: try $203 \times 203 \times 86$ kg/m UC. The relevant properties from section tables are as follows:

Flange thickness $T = 20.5\,\text{mm}$

Area $A_g = 110\,\text{cm}^2 = 110 \times 10^2\,\text{mm}^2$

Radius of gyration $r_x = 9.27\,\text{cm} = 92.7\,\text{mm}$

Radius of gyration $r_y = 5.32\,\text{cm} = 53.2\,\text{mm}$

It has already been stated that all UC sections when acting as columns are classified as semi-compact; therefore it is unnecessary to show that the section is not slender.

The ultimate design strength p_y for grade 43 steel sections, from Table 5.1, is $275\,\text{N/mm}^2$ provided that the flange thickness does not exceed 16 mm. If the flange thickness is greater than 16 mm then p_y reduces to $265\,\text{N/mm}^2$. In this case $T = 20.5\,\text{mm} > 16\,\text{mm}$, and therefore $p_y = 265\,\text{N/mm}^2$.

The slenderness values are given by

$$\lambda_x = \frac{L_{\text{Ex}}}{r_x} = \frac{5000}{92.7} = 54 < 180$$

$$\lambda_y = \frac{L_{\text{Ey}}}{r_y} = \frac{5000}{53.2} = 94 < 180$$

These are satisfactory.

The relevant BS 5950 strut table to use may be determined from Table 5.11. For buckling about the x–x axis use Table 27b; for buckling about the y–y axis use Table 27c. Hence

For $\lambda_x = 54$ and $p_y = 265\,\text{N/mm}^2$: $\quad p_c = 223\,\text{N/mm}^2$

For $\lambda_y = 94$ and $p_y = 265\,\text{N/mm}^2$: $\quad p_c = 133\,\text{N/mm}^2$

Therefore p_c for design is $133\,\text{N/mm}^2$.

The compression resistance is given by

$$P_c = A_g p_c = 110 \times 10^2 \times 133 = 1\,463\,000\,\text{N} = 1463\,\text{kN} > 1400\,\text{kN}$$

That is, $P_c > F$. Thus:

Adopt $203 \times 203 \times 86\,\text{kg/m UC}$.

The *Steelwork Design Guide to BS 5950* produced by the Steel Construction Institute contains tables giving resistances and capacities for grade 43 UCs subject to both axial load and bending. A typical example for a number of UC sections is reproduced here as Table 5.12. From the table it may be seen that for the $203 \times 203 \times 86\,\text{kg/m UC}$ section that has just been checked, the relevant axial capacity P_{cy} is given as 1460 kN. This again shows the advantage of such tables for reducing the amount of calculation needed to verify a section.

Example 5.12

If a tie beam were to be introduced at the mid-height of the column in Example 5.11, as shown in Figure 5.32, determine a suitable grade 43 UC section.

Ultimate axial load $F = 1400\,\text{kN}$

By introducing a tie at mid-height on either side of the y–y axis, the section is effectively pinned at mid-height and hence the effective height about the y–y axis

Table 5.12 Universal columns subject to axial load and bending, steel grade 43: compression resistance P_{cx}, P_{cy} (kN) and buckling resistance moment M_b (kN m) for effective length L_e (m), and reduced moment capacity M_{rx}, M_{ry} (kN m) for ratios of axial load to axial load capacity F/P_z (abstracted from the *Steelwork Design Guide to BS 5950 Part 1*, published by the Steel Construction Institute)

Designation and capacities	L_e(m) F/P_z	1.5 0.05	2.0 0.10	2.5 0.15	3.0 0.20	3.5 0.25	4.0 0.30	5.0 0.35	6.0 0.40	7.0 0.45	8.0 0.50	9.0 0.55	10.0 0.60	11.0 0.65
$203 \times 203 \times 86$	P_{cx}	2920	2870	2810	2750	2680	2610	2450	2260	2040	1810	1580	1370	1190
$P_z = 2920$	P_{cy}	2740	2570	2400	2220	2030	1830	1460	1150	916	740	607	0	0
$M_{cx} = 259$	M_b	259	259	259	253	244	235	220	206	194	182	172	163	155
$M_{cy} = 95$	M_{bs}	259	259	259	259	259	254	233	211	190	168	148	130	115
$p_y Z_y = 79$	M_{rx}	258	253	245	235	221	207	194	180	166	152	138	123	108
	M_{ry}	95	95	95	95	95	95	95	95	95	95	95	90	82
$203 \times 203 \times 71$	P_{cx}	2410	2380	2330	2270	2220	2160	2020	1860	1680	1480	1290	1120	970
$P_z = 2410$	P_{cy}	2260	2130	1980	1830	1670	1510	1200	943	750	605	496	0	0
$M_{cx} = 213$	M_b	213	213	213	203	195	187	173	160	149	139	131	123	116
$M_{cy} = 78$	M_{bs}	213	213	213	213	213	207	190	172	154	137	120	106	93
$p_y Z_y = 65$	M_{rx}	211	207	200	191	181	170	158	147	135	124	112	100	88
	M_{ry}	78	78	78	78	78	78	78	78	78	78	78	73	67
$203 \times 203 \times 60$	P_{cx}	2080	2040	2000	1950	1900	1850	1730	1580	1400	1230	1060	914	789
$P_z = 2080$	P_{cy}	1940	1820	1700	1560	1410	1270	995	778	615	494	404	0	0
$M_{cx} = 179$	M_b	179	179	176	168	160	152	138	126	116	107	99	92	86
$M_{cy} = 65$	M_{bs}	179	179	179	179	179	173	158	143	127	112	98	85	75
$p_y Z_y = 54$	M_{rx}	178	175	170	162	153	144	134	124	114	105	94	84	74
	M_{ry}	65	65	65	65	65	65	65	65	65	65	65	62	57
$203 \times 203 \times 52$	P_{cx}	1830	1790	1750	1710	1670	1620	1510	1370	1220	1070	921	792	683
$P_z = 1830$	P_{cy}	1700	1600	1480	1360	1230	1100	865	676	534	429	351	0	0
$M_{cx} = 156$	M_b	156	156	152	144	137	130	117	106	96	87	80	74	69
$M_{cy} = 57$	M_{bs}	156	156	156	156	156	150	137	124	110	96	84	74	64
$p_y Z_y = 47$	M_{rx}	155	152	148	141	133	125	116	108	99	90	81	73	64
	M_{ry}	57	57	57	57	57	57	57	57	57	57	57	54	49
$203 \times 203 \times 46$	P_{cx}	1620	1580	1550	1510	1470	1430	1330	1210	1070	934	804	691	593
$P_z = 1620$	P_{cy}	1500	1410	1310	1200	1080	968	757	590	465	374	305	0	0
$M_{cx} = 137$	M_b	137	137	132	125	118	111	99	88	79	72	66	60	56
$M_{cy} = 49$	M_{bs}	137	137	137	137	137	131	120	108	95	83	73	64	55
$p_y Z_y = 41$	M_{rx}	136	133	129	124	116	109	102	94	86	79	71	63	55
	M_{ry}	49	49	49	49	49	49	49	49	49	49	49	47	43
$152 \times 152 \times 37$	P_{cx}	1280	1240	1210	1160	1110	1060	928	783	647	532	441	369	312
$P_z = 1300$	P_{cy}	1140	1030	910	787	671	568	411	306	0	0	0	0	0
$M_{cx} = 85$	M_b	85	82	77	72	68	64	57	52	47	43	39	36	34
$M_{cy} = 30$	M_{bs}	85	85	85	82	77	72	62	52	44	37	31	26	22
$p_y Z_y = 25$	M_{rx}	84	83	81	77	73	68	64	59	54	50	45	40	35
	M_{ry}	30	30	30	30	30	30	30	30	30	30	30	29	26
$152 \times 152 \times 30$	P_{cx}	1030	1000	969	933	893	847	740	622	512	420	347	290	246
$P_z = 1050$	P_{cy}	914	825	727	627	533	450	325	241	0	0	0	0	0
$M_{cx} = 67$	M_b	67	64	60	56	52	48	42	37	33	30	27	25	23
$M_{cy} = 24$	M_{bs}	67	67	67	65	61	57	49	41	34	28	24	20	17
$p_y Z_y = 20$	M_{rx}	67	66	64	61	58	54	51	47	43	39	35	32	28
	M_{ry}	24	24	24	24	24	24	24	24	24	24	24	23	21
$152 \times 152 \times 23$	P_{cx}	802	777	751	722	688	650	560	465	379	310	255	213	180
$P_z = 820$	P_{cy}	706	632	552	472	397	334	239	177	0	0	0	0	0
$M_{cx} = 45$	M_b	50	47	43	39	36	33	28	24	21	19	17	15	14
$M_{cy} = 14$	M_{bs}	50	50	50	47	44	41	35	29	24	20	17	14	12
$p_y Z_y = 14$	M_{rx}	50	49	48	46	44	41	38	35	33	30	27	24	21
	M_{ry}	17	17	17	17	17	17	17	17	17	17	17	17	16

F is factored axial load.

M_b is obtained using an equivalent slenderness $= nuvL_e/r$ with $n = 1.0$.

M_{bs} is obtained using an equivalent slenderness $= 0.5L/r$.

Values have not been given for P_{cx} and P_{cy} if the values of slenderness are greater than 180.

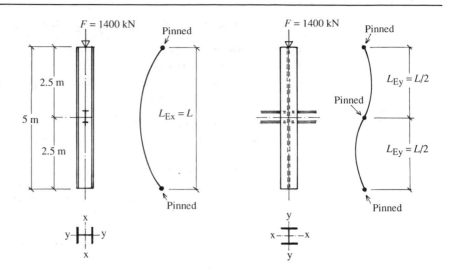

Figure 5.32   *Column load and effective lengths*

is halved. The effective height about the x–x axis will be unchanged. Thus

Effective length $L_{Ex} = 1.0L = 5000\,\text{mm}$

Effective length $L_{Ey} = \dfrac{L}{2} = \dfrac{5000}{2} = 2500\,\text{mm}$

It is again necessary to assume a trial section for checking: try $203 \times 203 \times 52\,\text{kg/m}$ UC. The relevant properties from section tables are as follows:

Flange thickness $T = 12.5\,\text{mm}$
Area $A_g = 66.4\,\text{cm}^2 = 66.4 \times 10^2\,\text{mm}^2$
Radius of gyration $r_x = 8.9\,\text{cm} = 89\,\text{mm}$
Radius of gyration $r_y = 5.16\,\text{cm} = 51.6\,\text{mm}$

Here $T = 12.5\,\text{mm} < 16\,\text{mm}$, and therefore $p_y = 275\,\text{N/mm}^2$. The slenderness values are given by

$$\lambda_x = \frac{L_{Ex}}{r_x} = \frac{5000}{89} = 56 < 180$$

$$\lambda_y = \frac{L_{Ey}}{r_y} = \frac{2500}{51.6} = 48 < 180$$

These are satisfactory.

The relevant strut tables to use, as determined from Table 5.11, are the same as in Example 5.11. Hence

For $\lambda_x = 56$ and $p_y = 275\,\text{N/mm}^2$: $p_c = 227\,\text{N/mm}^2$

For $\lambda_y = 48$ and $p_y = 275\,\text{N/mm}^2$: $p_c = 224\,\text{N/mm}^2$

It should be noted that even though the slenderness about the x–x axis is greater than that about the y–y axis, the lower value of p_c for λ_y will be the design criterion. Therefore p_c for design is $224\,\text{N/mm}^2$.

The compression resistance is given by

$$P_c = A_g p_c = 66.4 \times 10^2 \times 224 = 1\,487\,360\,\text{N} = 1487\,\text{kN} > 1400\,\text{kN}$$

That is, $P_c > F$. Thus:

Adopt $203 \times 203 \times 52\,\text{kg/m}$ UC.

The value of P_{cy} given in Table 5.12 for this section is $1480\,\text{kN}$.

It should be appreciated that the purpose of this example is not to advocate the introduction of tie beams in order to reduce the effective length of columns, but to illustrate the advantage of taking such beams or similar members into account when they are already present.

Figure 5.33 illustrates the effect on the slenderness that a similar mid-height tie would have if the column had been restrained in position and direction at both the cap and the base. The design procedure for the column would then be exactly the same as this example.

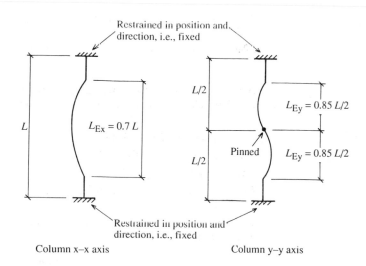

Figure 5.33   *Column effective lengths*

5.12.3 Axially loaded columns with nominal moments

The design of steel elements dealt with in this manual is based upon the principles of simple design. This assumes that the end connections between members are such that they cannot develop any significant restraint moments.

In practice the loads supported by columns are seldom applied concentrically, and therefore the effect of eccentric loading should be considered. For simple construction, where the end connections are not intended to transmit significant bending moments, the degree of eccentricity may be taken as follows:

(a) For a beam supported on a column cap plate, such as that shown in Figure 5.34a or b, the load is taken as acting at the face of the column.

(b) Where beams are connected by simple connections to the face of a column, as in Figure 5.35a and b, the load should be taken as acting at 100 mm from the column face, as shown in Figure 5.36.

(c) When a roof truss is supported on a column cap plate, as shown in Figure 5.37, and the connection cannot develop significant moments, the eccentricity may be neglected.

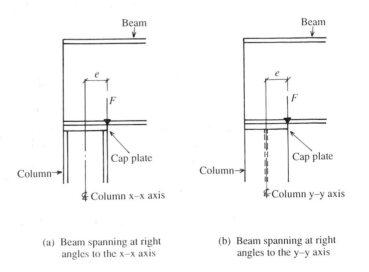

(a) Beam spanning at right angles to the x–x axis

(b) Beam spanning at right angles to the y–y axis

Figure 5.34 *Beams supported on a column cap plate*

A load applied eccentrically will induce a nominal bending moment in the column equal to the load times the eccentricity:

$$M_e = Fe$$

The effect on the column of this moment must be examined in conjunction with the axial load.

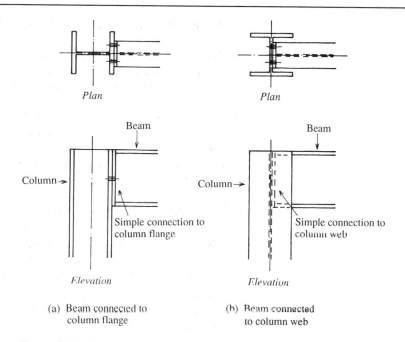

Plan   *Plan*

Beam   Beam

Column→   Column→

Simple connection to
column flange

Simple connection to
column web

Elevation   *Elevation*

(a) Beam connected to
column flange

(b) Beam connected
to column web

Figure 5.35 *Beams connected to the face of a column*

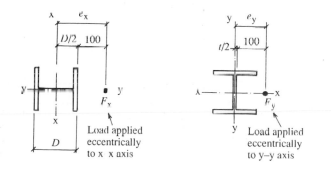

Figure 5.36 *Load eccentricity for beams connected to the face of a column*

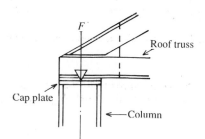

Figure 5.37 *Column supporting a roof truss, the load from which is transmitted concentrically*

Generally there are two separate checks that need to be applied to axially loaded columns with moments; they are a local capacity check and an overall buckling check.

Local capacity check

The local capacity of a column should be checked at the point of greatest bending moment and axial load. It will vary depending on the section classification, and therefore two relationships are given in BS 5950. One is for semi-compact and slender cross-sections, whilst the other is for plastic and compact cross-sections. For simplicity the relationship for semi-compact and slender cross-sections may be used to check all columns except those designed by plastic analysis. Therefore if elastic analysis is employed only one relationship need be satisfied, which is as follows:

$$\frac{F}{A_g p_y} + \frac{M_x}{M_{cx}} + \frac{M_y}{M_{cy}} \leqslant 1$$

where

F applied axial load

A_g gross sectional area, from section tables

p_y design strength of the steel

M_x applied moment about the major axis

M_{cx} moment capacity about the major axis in the absence of axial load; see Section 5.10.2 for beams

M_y applied moment about the minor axis

M_{cy} moment capacity about the minor axis in the absence of axial load; see Section 5.10.2 for beams

It should be noted that if the column section is classified as slender, the design strength p_y of the steel would be reduced. This does not apply to any UC sections since none is classed as slender.

Overall buckling check

A simplified approach and a more exact approach are offered in BS 5950 for the overall buckling check. The simplified approach must always be used to check columns subject to nominal moments. Since only columns with nominal moments are dealt with in this manual only the simplified approach will be considered here, for which the following relationship must be satisfied:

$$\frac{F}{A_g p_c} + \frac{mM_x}{M_b} + \frac{mM_y}{p_y Z_y} \leqslant 1$$

where

F applied axial load

A_g gross sectional area, from section tables

p_c compressive strength

m has value 1 when only nominal moments are applied

M_x applied moment about the major axis

M_b buckling resistance capacity about the major axis

M_y applied moment about the minor axis

p_y design strength of the steel

Z_y elastic section modulus about the minor axis, from section tables

It should be noted that when $m = 1$ the overall buckling check will always control the design. Therefore for columns supporting only nominal moments it is not necessary to carry out the local capacity check discussed in the previous section.

The buckling resistance capacity M_b of the section about the major axis is obtained from the following expression:

$$M_b = p_b S_x$$

where p_b is the bending strength and S_x is the plastic modulus of the section about the major axis, obtained from section tables. The bending strength for columns is obtained from BS 5950 Table 11, reproduced earlier as Table 5.5. It depends on the steel design strength p_y and the equivalent slenderness λ_{LT}, which for columns supporting only nominal moments may be taken as

$$\lambda_{LT} = 0.5 \frac{L}{r_y}$$

where L is the distance between levels at which both axes are restrained, and r_y is the radius of gyration of the section about its minor axis, from section tables.

5.12.4   Design summary for axially loaded steel columns with nominal moments

The procedure for the design of axially loaded columns with nominal moments, using grade 43 UC sections, may be summarized as follows:

(a) Calculate the ultimate axial load F applied to the column.

(b) Select a trial section.

(c) Calculate the nominal moments M_x and M_y about the respective axes of the column.

(d) Determine the overall effective length L_E from the guidance given in Table 5.10

(e) Calculate the slenderness λ from L_E/r and ensure that it is not greater than 180.

(f) Using the slenderness λ and the steel design strength p_y, obtain the compression strength p_c from Table 27a–d of BS 5950.

(g) Obtain the bending strength p_b from Table 5.5 using the steel design strength p_y and the equivalent slenderness λ_{LT}, which may be taken as $0.5 L/r_y$ for columns subject to nominal moments.

(h) Calculate M_b from the expression $M_b = p_b S_x$.

(i) Ensure that the following relationship is satisfied:

$$\frac{F}{A_g p_c} + \frac{m M_x}{M_b} + \frac{m M_y}{p_y Z_y} \leqslant 1$$

Example 5.13

Design a suitable grade 43 UC column to support the ultimate loads shown in Figure 5.38. The column is effectively held in position at both ends and restrained in direction at the base but not at the cap.

Ultimate axial load $F = 125 + 125 + 285 + 5 = 540\,\text{kN} = 540 \times 10^3\,\text{N}$

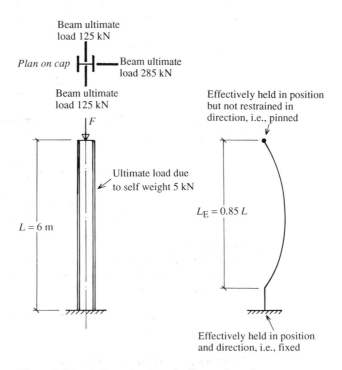

Figure 5.38 *Column loads and effective length*

Once again it is necessary to assume a trial section for checking: try $203 \times 203 \times 60\,\text{kg/m UC}$. The relevant properties from section tables are as follows:

Depth $D = 209.6\,\text{mm}$; width $B = 205.2\,\text{mm}$

Flange thickness $T = 14.2\,\text{mm}$

Web thickness $t = 9.3\,\text{mm}$

Area $A_g = 75.8\,\text{cm}^2 = 75.8 \times 10^2\,\text{mm}^2$

Radius of gyration $r_x = 8.96\,\text{cm} = 89.6\,\text{mm}$

Radius of gyration $r_y = 5.19\,\text{cm} = 51.9\,\text{mm}$

Plastic modulus $S_x = 652\,\text{cm}^3 = 652 \times 10^3\,\text{mm}^3$

It has already been stated that all UC sections are semi-compact, and therefore it is unnecessary to show that the section is not slender.

The eccentricity e_x is given by

$$e_x = \frac{D}{2} + 100 = \frac{209.6}{2} + 100 = 204.8\,\text{mm}$$

Then

$$\text{Nominal moment } M_x = \text{beam reaction} \times e_x$$
$$= 285 \times 204.8 = 58\,368\,\text{kN mm} = 58.368 \times 10^6\,\text{N mm}$$

Since the beam reactions are the same on either side of the y–y axis there will be no bending about this axis: therefore $M_y = 0$.

It is not necessary to check the local buckling capacity of columns subject to nominal moments. The overall buckling check using the simplified approach should be carried out to ensure that the following relationship is satisfied:

$$\frac{F}{A_g p_c} + \frac{mM_x}{M_b} + \frac{mM_y}{p_y Z_y} \leqslant 1$$

The compression strength p_c is calculated as follows. First, $T = 14.2\,\text{mm} < 16\,\text{mm}$. Therefore $p_y = 275\,\text{N/mm}^2$. The slenderness values are given by

$$\lambda_x = \frac{L_E}{r_x} = \frac{0.85L}{r_x} = \frac{0.85 \times 6000}{89.6} = 57 < 180$$

$$\lambda_y = \frac{L_E}{r_y} = \frac{0.85L}{r_y} = \frac{0.85 \times 6000}{51.9} = 98 < 180$$

These are satisfactory.

The relevant BS 5950 strut tables to use may be determined from Table 5.11. For buckling about the x–x axis use Table 27 b; for buckling about the y–y axis use Table 27 c. Hence

For $\lambda_x = 57$ and $p_y = 275\,\text{N/mm}^2$: $p_c = 225\,\text{N/mm}^2$

For $\lambda_y = 98$ and $p_y = 275\,\text{N/mm}^2$: $p_c = 129\,\text{N/mm}^2$

Therefore p_c for design is $129\,\text{N/mm}^2$.

For columns subject to nominal moments, m may be taken as 1.0.

The buckling resistance moment M_b for columns subject to nominal moments is calculated as follows. First,

$$\lambda_{LT} = \frac{0.5L}{r_y} = \frac{0.5 \times 6000}{51.9} = 58$$

Next, $p_b = 218 \, \text{N/mm}^2$ by interpolation from Table 5.5. Therefore

$$M_b = p_b S_x = 218 \times 652 \times 10^3 = 142.14 \times 10^6 \, \text{N mm}$$

Hence

$$\frac{F}{A_g p_c} + \frac{mM_x}{M_b} + \frac{mM_y}{p_y Z_y} = \frac{540 \times 10^3}{75.8 \times 10^2 \times 129} + \frac{1 \times 58.368 \times 10^6}{142.14 \times 10^6} + 0$$

$$= 0.55 + 0.41 + 0 = 0.96 < 1.0$$

Adopt $203 \times 203 \times 60 \, \text{kg/m}$ UC.

5.12.5 Cased columns

If steel columns are to be cased in concrete, for fire protection perhaps, structural advantage may be taken of the casing if certain requirements are met with respect to the concrete and reinforcement. The requirements in relation to UC sections are basically as follows:

(a) The steel section is unpainted and free from oil, grease, dirt or loose rust and millscale.

(b) The steel section is solidly encased in ordinary dense structural concrete of at least grade 30 to BS 8110.

(c) The surface and edges of the flanges of the steel section have a concrete cover of not less than 50 mm.

(d) The casing is reinforced using steel fabric, reference D 98, complying with BS 4483. Alternatively steel reinforcement not less than 5 mm diameter, complying with BS 4449 or BS 4482, may be used in the form of a cage of longitudinal bars held by closed links at a maximum spacing of 200 mm. The maximum lap of the reinforcement and the details of the links should comply with BS 8110.

(e) The reinforcement is so arranged as to pass through the centre of the concrete cover.

A typical cross-section through a cased UC satisfying these requirements is shown in Figure 5.39.

The allowable load for concrete cased columns is based upon certain empirical rules given in BS 5950. Those relating to axially loaded cased columns are as follows:

(a) The effective length L_E is limited to the least of $40b_c$ or $100b_c^2/d_c$ or $250r$, where

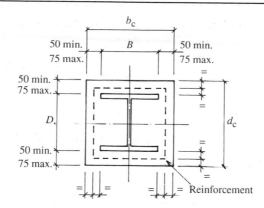

Figure 5.39   *Typical cross-section through a cased UC column*

b_c minimum width of solid casing within the depth of the steel section, as indicated in Figure 5.39

d_c minimum depth of solid casing within the width of the steel section, as indicated in Figure 5.39

r minimum radius of gyration of the uncased steel section, that is
r_y for UC sections

(b) The radius of gyration r_y of the cased section should be taken as $0.2b_c$
but never more than $0.2(b + 150)$ mm. This implies that any casing
above 75 mm cover should be ignored for structural purposes. The
radius of gyration r_x should be taken as that of the uncased steel
section.

(c) The compression resistance P_c of a cased column should be determined from the following expression:

$$P_c = \left(A_g + 0.45\frac{f_{cu}}{p_y} A_c \right) p_c$$

However, this should not be greater than the short strut capacity of
the section, given by

$$P_{cs} = \left(A_g + 0.25\frac{f_{cu}}{p_y} A_c \right) p_y$$

where

A_c gross sectional area of the concrete, ($b_c d_c$ in Figure 5.39) but
neglecting any casing in excess of 75 mm or any applied finish

A_g gross sectional area of the steel section

f_{cu} characteristic concrete cube strength at 28 days, which should
not be greater than 40 N/mm²

p_c compressive strength of the steel section determined in the manner described for uncased columns in Section 5.12.1, but using the r_y and r_x of the cased section

p_y design strength of the steel: $p_y \leqslant 355 \, \text{N/mm}^2$

P_{cs} short strut capacity, that is the compression resistance of a cased strut of zero slenderness

When a cased column is subject to axial load and bending it must satisfy the following relationships:

(a) Local capacity check:

$$\frac{F_c}{P_{cs}} + \frac{M_x}{M_{cx}} + \frac{M_y}{M_{cy}} \leqslant 1$$

(b) Overall buckling resistance:

$$\frac{F_c}{P_c} + \frac{mM_x}{M_b} + \frac{mM_y}{M_{cy}} \leqslant 1$$

The radius of gyration r_y for calculating the buckling resistance moment M_b of a cased column should be taken as the greater of the r_y of the uncased section or $0.2(B + 100) \, \text{mm}$, where B is as indicated in Figure 5.39. The value of M_b for the cased section must not exceed $1.5 M_b$ for the same section uncased.

Example 5.14

Determine the compression resistance of the grade 43 $203 \times 203 \times 86 \, \text{kg/m} \, \text{UC}$ column shown in Figure 5.40, which is structurally cased to the minimum requirements of BS 5950. The column is effectively held in position at both ends but not restrained in direction, as indicated in Figure 5.41

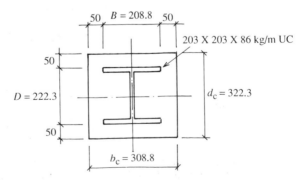

Figure 5.40   *Cross-section through cased column*

The properties of the cased section are as follows, from section tables where appropriate:

Gross area of concrete $A_c = b_c d_c = 308.8 \times 322.3 = 99\,526 \, \text{mm}$

Gross sectional area of steel section $A_g = 110 \, \text{cm}^2 = 110 \times 10^2 \, \text{mm}^2$

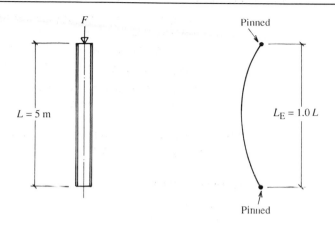

Figure 5.41 *Effective length of column*

$r_x = r_x$ of uncased section $= 9.27\,\text{cm} = 92.7\,\text{mm}$

r_y of uncased section $= 5.32\,\text{cm} = 53.2\,\text{mm}$

r_y of cased section $= 0.2b_c = 0.2 \times 308.8 = 61.8\,\text{mm}$

Effective length $L_E = L = 5000\,\text{mm}$

Check that the effective length does not exceed the limiting values for a cased column:

$$40b_c = 40 \times 308.8 = 12\,352\,\text{mm} > 5000\,\text{mm}$$

$$\frac{100b_c^2}{d_c} = \frac{100 \times 308.8^2}{322.3} = 29\,587\,\text{mm} > 5000\,\text{mm}$$

$$250r_y \text{ of uncased section} = 250 \times 53.2 = 13\,300 > 5000\,\text{mm}$$

Here $T = 20.5\,\text{mm} > 16\,\text{mm}$. Therefore $p_y = 265\,\text{N/mm}^2$. The slenderness values are given by

$$\lambda_x = \frac{L_E}{r_x} = \frac{5000}{92.7} = 54 < 180$$

$$\lambda_y = \frac{L_E}{r_y} = \frac{5000}{61.8} = 81 < 180$$

These are satisfactory.

The relevant BS 5950 strut tables to use may be determined from Table 5.11. For buckling about the x–x axis use Table 27b; for buckling about the y–y axis use Table 27c. Hence

For $\lambda_x = 54$ and $p_y = 265\,\text{N/mm}^2$: $\quad p_c = 223\,\text{N/mm}^2$

For $\lambda_y = 81$ and $p_y = 265\,\text{N/mm}^2$: $\quad p_c = 155\,\text{N/mm}^2$

Therefore p_c for design is $155\,\text{N/mm}^2$.

The compression resistance is given by

$$P_c = \left(A_g + 0.45\frac{f_{cu}}{p_y}A_c\right)p_c$$

$$= \left(110 \times 10^2 + 0.45 \times \frac{20}{265} \times 99\,526\right)155$$

$$= (11\,000 + 3380)155 = 14\,380 \times 155 = 2\,228\,900\,\text{N} = 2229\,\text{kN}$$

This must not be greater than the short strut capacity P_{cs} of the section, given by

$$P_{cs} = \left(A_g + 0.25\frac{f_{cu}}{p_y}A_c\right)p_y$$

$$= \left(11\,000 + 0.25 \times \frac{20}{265} \times 99\,526\right)265$$

$$= (11\,000 + 1878)265 = 12\,878 \times 265 = 3\,412\,670\,\text{N} = 3413\,\text{kN} > 2229\,\text{kN}$$

Therefore the compression resistance of the cased column is 2229 kN. This may be compared with the compression resistance of 1463 kN for the same section uncased that was calculated in Example 5.11. Thus the load capacity of the section when cased has increased by 52 per cent.

5.12.6   Column baseplates

The column designs contained in this manual relate to axially loaded columns and columns subject to nominal moments at the cap. Therefore only the design of baseplates subject to compressive loading will be included here.

Empirical rules are given in BS 5950 for the design of slab baseplates, as illustrated in Figure 5.42, when subject to compressive loads only. When a column is concentrically loaded it may be assumed that the load at the base is transmitted uniformly over the area of the steel baseplate to the foundation concrete.

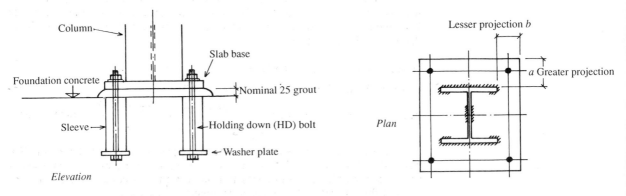

Elevation

Figure 5.42   *Typical slab base*

The bearing strength for concrete foundations may be taken as $0.4f_{cu}$, where f_{cu} is the characteristic concrete cube strength at 28 days as indicated in Table 5.13. This enables the area of baseplate to be calculated and suitable plan dimensions to be determined. The baseplate thickness is then determined from the following expression:

$$t = \left[\frac{2.5}{p_{yp}} w(a^2 - 0.3b^2) \right]^{1/2}$$

but t must not be less than the flange thickness of the column. In this expression,

a   greater projection of the plate beyond the column (see Figure 5.42)

b   lesser projection of the plate beyond the column (see Figure 5.42)

w   pressure on the underside of the plate assuming a uniform distribution (N/mm^2)

p_{yp} design strength of the plate, which may be taken as p_y given in Table 5.1, but not greater than 270 N/mm^2

Table 5.13   Bearing strength for concrete foundations

Concrete grade	Characteristic cube strength at 28 days f_{cu} (N/mm^2)	Bearing strength $0.4 f_{cu}$ (N/mm^2)
C30	30	12.0
C35	35	14.0
C40	40	16.0
C45	45	18.0
C50	50	20.0

Example 5.15

Design a suitable slab baseplate for a $203 \times 203 \times 86$ kg/m UC supporting an ultimate axial load of 1400 kN if the foundation is formed from grade 30 concrete. It should be noted that this is the column for which the steel section was originally designed in Example 5.11.

Grade 30 concrete $f_{cu} = 30$ N/mm^2

Bearing strength from Table 5.13 $= 12$ N/mm^2

$$\text{Area of slab baseplate required} = \frac{\text{axial load}}{\text{bearing strength}} = \frac{1400 \times 10^3}{12} = 116\,667 \text{ mm}^2$$

Since the column section is basically square, provide a square baseplate. The baseplate side $= \sqrt{(116\,667)} = 342$ mm. For practical reasons use a 350 mm square baseplate, for which the plan configuration taking into account the actual dimensions of the UC will be as indicated in Figure 5.43.

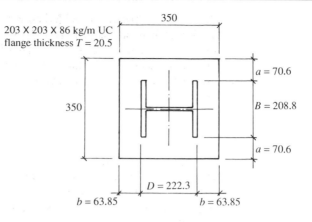

Figure 5.43   *Plan on baseplate*

The baseplate thickness is determined using the BS 5950 empirical expression:

$$t = \left[\frac{2.5}{p_{yp}} w(a^2 - 0.3b^2) \right]^{1/2}$$

where a and b are the dimensions shown in Figure 5.43. The bearing pressure is given by

$$w = \frac{1400 \times 10^3}{350 \times 350} = 11.43 \, \text{N/mm}^2$$

The design strength of the plate for grade 43 steel is obtained from Table 5.1, but must not be greater than $270 \, \text{N/mm}^2$. Since the flange thickness T of the UC column is 20.5 mm and the baseplate thickness t must not be less than this, the design strength from Table 5.1 will be $265 \, \text{N/mm}^2$. Therefore

$$t = \left[\frac{2.5}{265} \times 11.43(70.6^2 - 0.3 \times 63.85^2) \right]^{1/2}$$

$$= 20.14 \, \text{mm} < 20.5 \, \text{mm UC flange thickness}$$

Use a 25 mm thick baseplate. Thus finally:

Adopt a $350 \times 350 \times 25$ baseplate.

5.13   Connections

The design of connections usually follows the design of the principal components of a steel framed structure, and is normally regarded as part of the detailing process.

Connections may be bolted, welded or a combination of both. They must be proportioned with proper regard to the design method adopted for the structure as a whole. Therefore the bolts or welds making up a connection must be capable of transmitting all direct forces and resisting any bending moments.

The design of bolted or welded connections is beyond the scope of this manual, which is concerned with the design of individual elements. The British Constructional Steelwork Association publishes a book on the design of connections for joints in simple construction which would be useful for anyone with a particular interest in this topic. This and other sources of information relating to steel design are listed in the reference section.

5.14   References

BS 4 Structural steel sections.
　　Part 1 1980 Specification for hot-rolled sections.
BS 4360 1990 British Standard Specification for weldable structural steels.
BS 4848 Specification for hot-rolled structural steel sections.
　　Part 2 1991 Hollow sections.
　　Part 4 1972 Equal and unequal angles.
BS 5493 1977 Code of practice for protective coating of iron and steel structures against corrosion.
BS 5950 Structural use of steelwork in building.
　　Part 1 1990 Code of practice for design in simple and continuous construction: hot rolled sections.
　　Part 2 1985 Specification for materials, fabrication and erection: hot rolled sections.
Steelwork Design Guide to BS 5950: Part 1.
　　Volume 1 *Section Properties, Member Capacities* (1987).
　　Volume 2 *Worked Examples* (1986). Steel Construction Institute.
Introduction to Steelwork Design to BS 5950: Part 1. Steel Construction Institute, 1988.
Manual on Connections.
　　Volume 1 *Joints in Simple Construction Conforming with the Requirements of BS 5950: Part 1: 1985.* John W. Pask. British Constructional Steelwork Association, 1988.
Manual for the Design of Steelwork Building Structures. Institution of Structural Engineers, November 1989.

For further information contact:

The Steel Construction Institute, Silwood Park, Ascot, Berkshire, SL5 7QN.

The British Constructional Steelwork Association Ltd, 35 Old Queen Street, London, SW1H 9HZ.

Index